S0-BZS-660

Harvard Studies in East Asian Law

1. *Law in Imperial China: Exemplified by 190 Ch'ing Dynasty Cases (translated from the* Hsing-an hui-lan) *with Historical, Social, and Juridical Commentaries.* By Derk Bodde and Clarence Morris. 1967

2. *The Criminal Process in the People's Republic of China, 1949–1963: An Introduction.* By Jerome Alan Cohen. 1968

3. *Agreements of the People's Republic of China, 1949–1967: A Calendar.* By Douglas M. Johnston and Hungdah Chiu. 1968

4. *Contemporary Chinese Law: Research Problems and Perspectives.* Edited by Jerome Alan Cohen. 1970

5. *The People's Republic of China and the Law of Treaties.* By Hungdah Chiu. 1972

Harvard Studies in East Asian Law 5

The People's Republic of China and the Law of Treaties

The Harvard Law School, in cooperation with Harvard's East Asian Research Center, the Harvard-Yenching Institute, and scholars from other institutions, has initiated a program of training and research designed to further scholarly understanding of the legal systems of China, Japan, Korea, and adjacent areas. Accordingly, Harvard University Press has established a new series to include scholarly works on these subjects. The editorial committee consists of Jerome Alan Cohen (chairman), John K. Fairbank, L.S. Yang, and Donald Shively.

The People's Republic of China and the Law of Treaties

Hungdah Chiu

Harvard University Press
Cambridge, Massachusetts 1972

Library of Congress Catalog Card Number 72-173411
SBN 674-66175-3

Printed in the United States of America

To Yuan-yuan

Contents

Abbreviations

AJIL	*American Journal of International Law*
BFSP	British and Foreign State Papers
BYIL	*British Year Book of International Law*
CDSP	*Current Digest of the Soviet Press*
CFYC	*Cheng-fa yen-chiu* (Studies in political science and law)
CHYYC	*Chiao-hsuëh yü yen-chiu* (Teaching and research)
Cmd.	British Command Paper
DIA	Documents on International Affairs
DSB	*Department of State Bulletin*
EAS	Executive Agreement Series
Enlightenment Daily	Kuang-ming jih pao (Peking)
FH	*Fa-hsüeh* (Science of law)
FKHP	*Chung-hua jen-min kung-ho kuo fa-kuei hui-pien* (Collection of laws and regulations of the People's Republic of China), 13 vols.
FLHP	*Chung-yang jen-min cheng-fu fa-ling hui-pien* (Collection of laws and decrees of the Central People's Government)
ICJ	*International Court of Justice (Reports)*
ILM	*International Legal Materials*
KCWTYC	*Kuo-chi wen-t'i yen-chiu* (Studies in international problems)
LNTS	*League of Nations Treaty Series*
NCNA	New China News Agency or Hsinhua News Agency (Peking)
PC	*People's China*
People's Daily	*Jen-min jih-pao* (Peking)
PR	*Peking Review*
SCMP	*Survey of the China Mainland Press*
Stat.	United States Statutes at Large
TIAS	Treaties and Other International Acts Series
TS	British Foreign Office *Treaty Series*, 1957–1962
TYC	*Chung-hua jen-min kung-ho kuo t'iao-yüeh chi* (Compilation of treaties of the People's Republic of China)
UNTS	*United Nations Treaty Series*

UST — *United States Treaties and Other International Agreements*

WCC — *Chung-hua jen-min kung-ho kuo tui-wai kuan-hsi wen-chien chi* (Compilation of documents relating to the foreign relations of the People's Republic of China)

Preface

On May 22, 1969, the United Nations Conference on the Law of Treaties at Vienna adopted a new Convention on the Law of Treaties.[1] The Conference was attended by 110 states, nearly all of the existing states of the world. The universality of the Convention, however, is limited by the fact that Communist China (the People's Republic of China) could not participate in the Conference and has expressed no interest in becoming a party to the Convention.[2] In the future, the universality of the Convention will, to a significant extent, depend upon Communist China's willingness to subscribe to a law of treaties accepted by the vast majority of states of the world and their willingness to permit her to subscribe.

Before the establishment of the Communist regime, China concluded several hundred treaties or agreements with foreign states.[3] Since 1949 the People's Republic of China has concluded nearly 2,000 treaties or agreements, in a broad sense, with more than 70 states or political entities.[4] So long as Communist China does not accede to the Vienna Convention, it will be of paramount importance to the many states that find it necessary to deal with this world power to know its theory and practice regarding the law of treaties. Even if Communist China should adhere to the Vienna Convention, it would be important for states dealing with it to know its previous doctrines and practice in order to understand how the Convention may be interpreted and applied.

In recent years, a number of states have accused Communist China of treaty violations. For example, on August 11, 1958, the United States De-

1. U.N. Doc. A/conf. 39/27, May 23; reprinted in *AJIL*, 63:875 (1969) and *ILM*, 8:679, no. 4 (July 1969).

2. China was represented at the Conference by the government of the Republic of China. The Convention is open for signature or accession by all states that are members of the U.N., of any of the specialized agencies, or of the International Atomic Agency, or that are parties to the Statute of the International Court of Justice or that have been invited by the U.N. General Assembly to become parties to the Convention (Convention arts. 81, 83). China has been represented in the above-mentioned organizations by the government of the Republic of China, and the U.N. General Assembly has not invited Communist China to become a party to the Convention. So far, Communist China has ignored the Convention and no news media there have reported information concerning it.

3. For a collection of these treaties and agreements, see Inspector General of Customs, *Treaties, Conventions* [etc.] *between China and Foreign States* (Shanghai, 1917); MacMurray, ed., *Treaties and Agreements with and concerning China, 1894–1919* (New York, 1921); *Treaties and Agreements with and concerning China, 1919–1929* (Washington, 1929); and Ministry of Foreign Affairs, *Treaties between the Republic of China and Foreign States 1927–1957* (Taipei, 1958).

4. See Johnston and Chiu, *Agreements of the People's Republic of China, 1949–1967* (Cambridge, Mass., 1968).

partment of State issued a memorandum on recognition of Communist China to its missions abroad which stated that Communist China "has shown no intention to honor its international obligations [and] one of its first acts was to abrogate the treaties of the Republic of China, except those it chose to continue." It further stated that Communist China "has failed to honor various commitments entered into since [the establishment of the regime]."[5] Similarly, in a note to Communist China dated February 9, 1967, the Soviet Union charged that Communist China's anti-Soviet activities were "breaching the most elementary norms of international law and morality, not to mention the relations ensuing from the [1950 Sino-Soviet] Treaty on Friendship, Alliance and Mutual Aid."[6] On the other hand, Communist China has also frequently charged the United States, the Soviet Union, and other states with violating international agreements.

Accusations of this nature cannot be evaluated without first studying the fundamental problem of whether both sides subscribe to the same law of treaties. The purpose of this book is to analyze the theory and practice of the law of treaties followed by Communist China in the past twenty years, 1949 through 1969. It is based primarily on Communist Chinese sources, which may be divided into three general groups: (1) texts of treaties or agreements; (2) official statements and documents such as government announcements and diplomatic notes, and other authoritative statements such as *People's Daily* editorials; and (3) works of Communist Chinese writers.

This book seeks to be both analytical and comparative. Whenever appropriate, Communist Chinese doctrine and practice are compared with relevant Soviet or Western doctrine and practice and the provisions of the Vienna Convention. The book does not attempt to condemn or to defend the position of Communist China with respect to various problems of the law of treaties. It is simply an attempt to study Communist China's treaty law and practice in an objective way for the purpose of understanding and clarifying the subject.

The book was prepared under a contract with the United States Arms Control and Disarmament Agency. The judgments expressed are my own, and they do not necessarily reflect the views of the United States Arms Control and Disarmament Agency or any other organ of the United States government.

All translations not otherwise credited are my own.

5. "U.S. Policy on Non-Recognition of Communist China," *DSB,* 39:388, no. 1002 (Sept. 8, 1958).
6. *Pravda*, Feb. 10, 1967; trans. in *CDSP*, 19:6, no. 6 (Mar. 1, 1967).

I wish to express sincere thanks to Professor Jerome A. Cohen, of the Harvard Law School, for having read the manuscript and offered many valuable suggestions. I am grateful too to Professor Richard R. Baxter, also of Harvard Law School, for the experience and training I received in his seminar on the law of treaties. Professors Randle R. Edwards (Boston University), S. C. Leng (University of Virginia), and William E. Butler (London University) have read parts of the manuscript in its various stages of development and offered many constructive criticisms. I am grateful to them for valuable advice.

Finally, I am indebted to Ellen Tolstuk for ably typing and retyping my manuscript, to Mrs. Bertha Ezell for typing the final version, and to Lois Dougan Tretiak for general assistance.

Hungdah Chiu
Taipei, Taiwan, China
June 1, 1971

The People's Republic of China and the Law of Treaties

I The Position of Treaties in International Law and Municipal Law

The term "source of law" as used by Western students of international law is "the name for an historical fact out of which rules of conduct come into existence and legal force."[1] According to Oppenheim, the concept of "source" should not be confused with that of "cause." He gives the following illustration to explain the difference between source and cause:

> Source means a spring or well, and has to be defined as the rising from the ground of a stream of water. When we see a stream of water and want to know whence it comes, we follow the stream upwards until we come to the spot where it rises naturally from the ground. On that spot, we say, is the source of the stream of water. We know very well that this source is not the cause of the existence of the stream of water. Source signifies only the natural rising of water from a certain spot on the ground, whatever natural causes there may be for that rising (pp. 24–25).

Western writers are in general agreement that treaties and custom are the two principal sources of international law.[2] This is recognized by Article 38 of the Statute of International Court of Justice, which provides that the Court shall apply: "(a) International conventions, whether general or particular, establishing rules expressly recognized by the contesting states. (b) International custom, as evidence of a general practice accepted as law."

The Statute also recognizes a third separate source of law, namely, "the general principles of law recognized by civilized nations." In practice, however, the Court has seldom found occasion to apply "general principles of law," since as a rule conventional and customary international law has been deemed sufficient to supply the necessary basis for decision. There does not appear to be any consensus among Western writers concerning the meaning of these "general principles." The majority view seems to suggest that they are to be derived by selecting concepts common to all systems of municipal law.[3]

Judicial decisions and writings of publicists are considered by Western

1. Oppenheim, *International Law*, I, (8th ed. London, 1955), 25.
2. *Ibid.*, p. 29.
3. *Ibid.* See also O'Connell, *International Law,* I, (London, 1965), 10–11; Starke, *Introduction to International Law* (London, 1967), p. 31.

writers as subsidiary sources of international law; this is recognized by Article 38. In recent years, there has also been an increasing tendency to recognize the decisions or resolutions of international organizations as a separate source of international law.[4]

Oppenheim's distinction between source and cause has been criticized by a Communist Chinese writer, Ying T'ao, as "a circumlocution around terminology that avoids touching on the deeper social causes." In his view, "the substantive sources of bourgeois international law are the external policy of the bourgeoisie, which is also the will of the ruling class of those big capitalist powers."[5] He asserts that the so-called treaties, customs, and other "sources" are only the form in which international law is expressed; they can only be called *formal* sources. The emphasis on formal sources by bourgeois writers, he explains, is an intentional measure to deceive the people into believing that international law does not possess class character.

With respect to the formal sources of international law, the exact attitude of Communist China is not clear. In a book of selected documents on international law published in Peking in 1958 and edited by the Office for Teaching and Research of International Law of the Institute of Diplomacy, three documents were placed under the heading "sources of international law": (1) Article 38 of the Statute of the International Court of Justice; (2) Resolution of the United Nations General Assembly on Progressive Development of International Law and Its Codification (December 11, 1946); and (3) Resolution of the United Nations General Assembly on Affirmation of the Principles of International Law Recognized by the Charter of the Nürnberg Tribunal (December 11, 1946).[6]

4. See Tammes, "Decisions of International Organs as a Source of International Law," in academie de Droit International, *Recueil des cours*, 94:265 (Leyden, 1959); Starke, *Introduction to International Law*, pp. 50–52; O'Connell, *International Law*, I, 26–27; Fenwick, *International Law* (New York, 1965), pp. 96–97. For a discussion of Western views on the sources of international law, see Parry, *The Sources and Evidences of International Law* (Manchester, 1966).

5. Ying T'ao, "Recognize the True Face of Bourgeois International Law from a Few Basic Concepts," *KCWTYC*, 1:46, 47 (1960).

6. *Kuo-chi kung-fa ts'an-k'ao wen-chien hsüan-chi* (Selected reference documents of public international law) (Peking, 1958), pp. 11–12. This book apparently followed the pattern of a similar Soviet book published in 1957 and containing the following items under the heading "Sources of International Law": (1) art. 38, Statute of the International Court of Justice; (2) resolutions of the U.N. General Assembly on the progressive development of international law and its codification (Dec. 11, 1946); (3) decision concerning establishment of the U.N. International Law Commission (Nov. 12, 1947); (4a) organization of the International Law Commission as accepted on Dec. 18, 1956; (4b) functions of the Commission; (4c) cooperation of the Com-

But these materials hardly furnish conclusive evidence of Communist China's attitude toward the sources of international law. The editor of the above collection of documents makes it clear in his note that all documents included in the book are for "reference" only. Nevertheless, Communist China's recognition of treaties and customs as important sources seems beyond doubt. For example, Communist Chinese scholar Wei Liang writes that "treaties are an important source of international law and an important form of expressing international law."[7] And Communist China has frequently cited treaties as setting forth legal rules governing relations between states.[8] The frequent references to custom in diplomatic notes,[9] treaties,[10] and writings of Communist Chinese scholars[11] are a further indication that custom constitutes an important source of international law in Communist Chinese legal theory and practice.

In practice, Communist China's great interest in incorporating in many treaties, joint communiqués, and declarations the five principles of peaceful coexistence that are supposed to guide inter-state relations[12] indicates

---

mission with other organs. See Modzhorian and Sobakin, *Mezhdunarodnoe pravo v izbrannykh dokumentakh* (Moscow, 1957), pp. 12–21; Triska and Slusser, *The Theory, Law, and Policy of Soviet Treaties* (Stanford, 1962), pp. 30–31.

7. "On the Post Second World War International Treaties," in *Kuo-chi t'iao-yüeh chi* (International treaty series), 1953–1955, (Peking, 1961), p. 660.

8. *E.g.*, Communist China protested against Ghana for alleged unilateral tearing up of Sino-Ghanaian Agreement on Economic and Technical Cooperation, signed Aug. 18, 1961, *TYC*, 10:250; English trans. in "Sino-Ghanaian Economic and Technical Cooperation Agreement," *SCMP*, 2567:33–34 (Aug. 28, 1961). For the Chinese note of protest, see "China Strongly Protest Against Worsening of Sino-Ghanaian Relations by Ghanaian Authorities," *PR*, 9:8–10, no. 13 (Mar. 25, 1966).

9. *E.g.*, in the "Statement by the Chinese Embassy in Indonesia on the Forcible House Arrest of the Chinese Consul," May 13, 1960, it was asserted that "the forcible house arrest of Consul Chiang Yen, the crude encroachment upon the functions and rights, the personal safety and freedom of the consul . . . have violated the universally acknowledged international norms," *WCC*, 72:152; English trans. in "Statement by Chinese Embassy . . . " *PR*, 3:35, no. 20 (May 17, 1960)

10. *E.g.*, art. 2 of Sino-Indonesian Treaty of Friendship, signed Apr. 1, 1961, provides: "The Contracting Parties agree to continuously consolidate the diplomatic and consular relations between the two countries in accordance with the principle of reciprocity and international practice," *TYC*, 10:7; English trans. in "Treaty of Friendship between the People's Republic of China and the Republic of Indonesia" *PR*, 4:11, no. 24 (June 16, 1961).

11. *E.g.*, see Chou Keng-sheng, "The Persecution of Chinese Personnel by Brazilian Coup d'Etat Authority Is a Serious International Illegal Act," *People's Daily*, Apr. 24, 1964, p. 4. Chou argues that the persecution violated "international custom."

12. The five principles are: (1) mutual respect for each other's territorial integrity and sovereignty; (2) mutual nonaggression; (3) mutual noninterference in each other's internal affairs; (4) equality and mutual benefit; (5) peaceful coexistence. The principles were first included in the preamble of the Agreement between Communist

that it places great emphasis on the role of treaties as a source of international law. Some writers in Communist China believe that the growing number of treaties containing the five principles has transformed them into principles of general international law,[13] even though the principles are by no means novel and seem already to have been accepted by most states as principles of international law.[14] Moreover, a Communist Chinese writer points out that, although customs and treaties are equally binding under international law, treaties are nevertheless "more explicit" and will serve the function of eliminating arguments that may be used to avoid obligations under customary rules of international law.[15]

China and India on Trade and Intercourse between the Tibetan Region of China and India, April 29, 1954, *TYC*, 3:1; *UNTS*, 299:57.

The principles were explicitly mentioned in ten friendship treaties: Burma, Jan. 29, 1960 (Preamble), *TYC*, 9:44, English text in "Treaty of Friendship and Mutual Non-Aggression between the People's Republic of China and the Union of Burma," *PR*, 3:13, no. 5 (Feb. 2, 1960); Nepal, Apr. 28, 1960 (Preamble), *TYC*, 10:13, English text in "Sino-Nepalese Treaty of Peace and Friendship," *PR*, 3:6, no. 18 (May 3, 1960); Cambodia, Dec. 19, 1960 (Preamble), *TYC*, 9:25, English text in "Sino-Cambodian Treaty of Friendship and Mutual Non-Aggression," *PR*, 4:10, no. 19 (May 5, 1961); Indonesia, Apr. 1, 1961 (Preamble), *TYC*, 10:7, English text in "Treaty of Friendship between the People's Republic of China and the Republic of Indonesia," *PR*, 4:11, no. 24 (June 16, 1961); Yemen, June 9, 1964 (art. 2), *TYC*, 13:5, English text in "Sino-Yemeni Friendship Treaty," *PR*, 7:10, no. 25 (June 19, 1964); Guinea, Sept. 13, 1960 (art. 2), *TYC*, 10:17, English text in "Treaty of Friendship," *PR*, 4:7, no. 34 (Aug. 25, 1961); Congo(B), Oct. 2, 1964 (Preamble), *TYC*, 13:27, English text in "Sino-Congolese(B) Friendship Treaty," *PR*, 8:18, no. 3 (Jan. 15, 1965); Mali, Nov. 3, 1964 (art. 2), *TYC*, 13:38, English text in "China-Mali Friendship Treaty," *PR*, 8:25, no. 18 (Apr. 30, 1965); Tanzania, Feb. 20, 1965 (art. 2), *People's Daily*, Feb. 21, 1965, p. 1, English text in "Sino-Tanzanian Treaty of Friendship," *PR*, 8:9, no. 9 (Feb. 26, 1965).

The principles were mentioned in many other treaties and joint communiqués. Moreover, many treaties concluded by Communist China refer to one or more principles of peaceful coexistence even though they do not explicitly use the term "five principles of peaceful coexistence."

13. *E.g.*, Chao Yüeh: "In many important treaties concluded with . . . other countries, our country put forward a series of democratic principles which have had great influence and have become commonly observed principles among various countries," in "A Preliminary Criticism of Bourgeois International Law," *KCWTYC*, 3:8 (1959). See also Ting Ku, "Firmly Maintain the Five Principles of Peaceful Coexistence," *KCWTYC*, 2:2 (1959).

14. See Starke, *Introduction to International Law*, pp. 106–107. Even Chou Keng-sheng, a prominent Communist Chinese jurist, expressed a similar view in "The Principles of Peaceful Coexistence from the Viewpoint of International Law," *CFYC*, 6:37 (1955).

15. Chu Li-sun: "According to . . . custom . . . the use of nuclear weapons is illegal. Nonetheless, it is still of great significance to make efforts to conclude special conventions prohibiting nuclear weapons . . . Although general principles, customs, agreements and treaties are all equally binding under international law, nevertheless special conventions and agreements are more explicit. Therefore, concluding a special convention to prohibit nuclear weapons would undoubtedly be

The Communist Chinese position on the significance of treaties appears generally similar to the Soviet position; in the Soviet Union, writers generally emphasize them as the most important source of international law.[16] The origins of the Soviet position are understandable: as the first Communist country in the world, it had to emphasize treaties as the most important source or subject itself to a large number of existing norms and customs which had been formulated without its participation and which might have been detrimental to its interests. This concern may be less obvious in the case of Communist China because at the time of the establishment of the regime in 1949 a number of Communist countries already existed and had developed certain customs and norms in relations among themselves. But Communist China may still find it expedient to emphasize treaties in order to forestall the invocation by Western states of their own inter-state norms and customs against China. Moreover, the recent deterioration of diplomatic relations between Communist China and "revisionist" Communist countries would tend to strengthen its emphasis on the role of treaties, since now even customs and norms developed among other Communist countries may not be acceptable to Communist China.

Only treaties which are "just" or "equal" can be regarded as sources of international law, according to Chinese Communist writers. In this connection, Ying T'ao severely criticizes Western international law scholars for failing to appraise the substance of the treaties which they consider to be a source of international law:

> Since bourgeois international law scholars state that treaties are the principal source of international law, may we ask where those treaties come from? By referring to the large number of treaties concluded in the period of capitalism, it may be proved that these treaties were all concluded under the guidance of the external policy of capitalist countries in accordance with the demands of the bourgeoisie and through diplomatic means and arbitrary external practices. For instance, bourgeois international law refers to such documents as "the Covenant of the League of Nations" as an important source of interna-

---

an important step forward. It would serve the function of thoroughly refuting all arbitrary arguments of American atomic war mongers," in "The Use of Atomic and Hydrogen Weapons Is the Most Serious Criminal Act in Violation of International Law," *CFYC*, 4:33 (1955). See also Ch'en T'i-ch'iang, "The Illegality of Atomic Weapons from the Viewpoint of International Law," *Shih-chieh chih-shih* (World Knowledge), 4:12 (Feb. 20, 1955).

16. See Triska and Slusser, *The Theory, Law and Policy of Soviet Treaties*, pp. 9–31.

> tional law. But the "League of Nations" was merely a tool in the hands of imperialism, and the "Covenant" was signed by the victorious powers of the First World War to maintain acquired interests and the kind of international order they needed, and to facilitate their further suppression of colonies. Bourgeois international law also considers unequal treaties imposed upon weak and small countries to be sources of international law. The implications of this are only too obvious to everyone.[17]

The question of the position of treaties in municipal law is an important part of the broader problem of the relationship between international and municipal law. However, Communist Chinese writers rarely discuss either this specific question or the broader relationship; and those who do, such as Ying T'ao, confine themselves almost exclusively to criticisms of Western theories and practices. The 1954 Constitution of the People's Republic of China also is silent in this respect.

In practice, some rules of international law have been given legal effect in Communist China's municipal law. For example, Article 17 of the 1964 Act Regulating the Entry, Exit, Transit, Residence, and Travel of Foreign Nationals provides that "cases of violations . . . involving foreign nationals who enjoy diplomatic immunity shall be handled through diplomatic channels."[18] Moreover, it is interesting to note that in sentencing Japanese war criminals in 1956 the Supreme People's Court invoked "international law standards and humanitarian principles."[19]

Also, some treaties are included in the official publication entitled Collection of Laws and Regulations of the People's Republic of China (*FKHP*). Presumably these treaties have the same legal status as legislation included in the Collection.[20]

17. Ying T'ao, "Recognize the True Face," p. 47. See also Chao Yüeh, "A Preliminary Criticism," p. 3. Chao also criticized bourgeois international law for including many "unequal treaties" as sources of international law. Wang Yao-t'ien wrote that "unequal treaties are in violation of international law and do not have legal validity," *Kuo-chi mao-yi t'iao-yüeh ho hsieh-ting* (International trade treaties and agreements) (Peking, 1958), p. 10. Since unequal treaties are invalid, they certainly cannot be considered a source of international law.

18. *People's Daily*, Apr. 20, 1964, p. 3.

19. *Jih-pen wen-t'i wen-chien hui-pien* (Collection of documents on the problem of Japan), II, (Peking, 1958), 126.

20. The following treaties concluded between 1955 and 1963 were included in the official *Collection*: Nepal (preservation of friendly relations, commerce, and communication between the Tibetan Region of China and Nepal), Sept. 20, 1956, *TYC*, 5:4, *FKHP*, 4:231; East Germany (consular affairs), Jan. 27, 1959, *TYC*, 8:26, *FKHP*, 10:218; Soviet Union (consular affairs), June 23, 1959, *TYC*, 8:20, *FKHP*, 10:214; Indonesia (dual nationality), Apr. 22, 1955, *TYC*, 8:12, *FKHP*,

In order to implement a treaty in municipal law states sometimes have to enact laws or decrees. This situation exists in Communist China. For example, in 1961 a decree was issued by Communist China to implement the 1955 Sino-Indonesian Dual Nationality Treaty.[21]

11:100; Indonesia (exchange of notes re implementation of dual nationality treaty), June 3, 1955, *TYC*, 8:17, *FKHP*, 11:105; Nepal (boundary question), Mar. 21, 1960, *TYC*, 9:63, *FKHP*, 11:109; Burma (boundary question), Jan. 28, 1960, *TYC*, 9:65, *FKHP*, 11:111; East Germany (commerce and navigation), Jan. 18, 1960, *TYC*, 9:134, *FKHP*, 11:114; North Korea (cooperation in navigation and transportation on boundary rivers), May 23, 1960, *TYC*, 9:153, *FKHP*, 12:27; Indonesia (measures for implementation of dual nationality treaty), Dec. 15, 1960, *TYC*, 9:58, *FKHP*, 12:30; Burma (boundary), Oct. 1, 1960, *TYC*, 9:68, *FKHP*, 12:36; Czechoslovakia (consular affairs), May 7, 1960, *TYC*, 9:52, *FKHP*, 12:49; Albania (commerce and navigation), Feb. 2, 1961, *TYC*, 10:361, *FKHP*, 12:54; Nepal (boundary), Oct. 5, 1961, *TYC*, 10:45, *FKHP*, 12:59; Mongolia (commerce), Apr. 26, 1961, *TYC*, 10:361, *FKHP*, 13:76; Nepal (exchange of notes on questions of [1] selection of nationality and [2] crossing of boundary by inhabitants of boundary regions to till land and graze), Aug. 14, 1962, *TYC*, 11:15, *FKHP*, 13:82; North Korea (commerce and navigation), Nov. 5, 1962, *TYC*, 11:92, *FKHP*, 13:88; North Vietnam (commerce and navigation), Nov. 5, 1962, *TYC*, 11:100, *FKHP*, 13:93; Mongolia (boundary), Dec. 26, 1962, *TYC*, 11:19, *FKHP*, 13:98; Pakistan (boundary), Mar. 2, 1963, *TYC*, 12:64, *FKHP*, 13:118; Afghanistan (boundary), Nov. 22, 1963, *TYC*, 12:122, *FKHP*, 13:121.

21. "Provisions of the Ministry of Public Security governing the procedure to be followed by persons possessing simultaneously the nationalities of the People's Republic of China and the Republic of Indonesia in making a declaration selecting the nationality of the People's Republic of China" were promulgated Apr. 25, 1961, *FKHP*, 12:35.

II The Nature and Scope of Treaties

Communist Chinese writers rarely define the term "treaty"; the only two exceptions known to me are Wei Liang and Wang Yao-t'ien. Wei Liang defines a treaty as "an agreement between two or more states, which must have received the unanimous consent of all the contracting parties."[1] Wang Yao-t'ien's definition of a treaty is slightly different. An "international treaty" to him is "a document between two or more states concerning the establishment, change, or termination of their supreme rights and duties."[2]

Neither definition tells much about the Communist Chinese view of the nature and scope of treaties. The definitions do not, for instance, differentiate equal from unequal treaties – an important distinction to the Communist Chinese government and Chinese writers. For this reason, a somewhat detailed analysis of the nature and scope of treaties in Communist China's theory and practice, including a discussion of the various elements of treaties such as parties, form, and classification, is necessary. The problem of "semiofficial agreements" which, even though concluded by nongovernmental organs, serve official functions as treaties, must also be treated.

Parties to Treaties

The definitions of Wei Liang and Wang Yao-t'ien that a treaty is an agreement or document between two or more states seem to indicate that dependent entities, as well as international organizations, cannot become parties to treaties; moreover, the definitions do not take into account the problem of the capacity of revolutionary groups to become parties to treaties. These definitions are in fact too narrow to embrace Communist Chinese theory and practice concerning treaties.

Western jurists generally agree that at the present stage of international law international organizations can be parties to a treaty.[3] Lauterpacht's

1. Wei Liang, "Looking at the So-Called McMahon Line from the Angle of International Law," *KCWTYC*, 6:46–47 (1959).

2. Wang Yao-t'ien, *Kuo-chi mao-yi t'iao-yüeh ho hsieh-ting* (International trade treaties and agreements) (Peking, 1958), p. 9. Another writer who might have defined the terms "treaty" and "agreement" is Kuo Chao; see his "The Names and Kinds of International Treaties," *Chi-lin jih-pao* (Kirin daily), April 4, 1957, p. 4, which has not been available to me.

3. Although art. 1 of the 1969 Convention on the Law of Treaties provides that the Convention applies to treaties concluded by states only, it does not imply that

Oppenheim sees treaties as "agreements, of a contractual character, between States, or *organizations of States*, creating legal rights and obligations between the Parties"[4] (emphasis added). Some writers have further reasoned that since international organizations have the capacity to conclude treaties they are therefore subjects of international law.[5]

Communist Chinese theory, however, denies that international organizations are subjects of international law and maintains that these organizations do not have the capacity to conclude treaties in their own right. As for existing treaties concluded in the name of an international organization, a Communist Chinese writer explains that the true parties to such treaties are the members of that organization.[6]

The attitude of Communist Chinese writers on this subject conforms to the traditional Soviet theory concerning subjects of international law.[7]

---

international organizations should be excluded from becoming parties to treaties. The reason for placing this limitation upon the scope of the Convention was explained in the report of the United Nations International Law Commission, which prepared the draft of the Convention: "Treaties concluded by international organizations have many special characteristics; and the Commission considered that it would both unduly complicate and delay the drafting of the present articles if it were to attempt to include in them satisfactory provisions concerning treaties of international organizations," commentary to art. 1, Draft Articles on the Law of Treaties, Report of International Law Commission on 2nd Part of 17th Session, Jan. 3–28, 1966, and 18th Session, May 4–July 19, 1966, in U.N. General Assembly, *Official Records*, 21st sess., supp. no. 9 (U.N. Doc. A/6309/rev. 1), p. 20.

4. Oppenheim, *International Law*, I, 8th ed. (London, 1955), 877. For a study of this question see Chiu, *The Capacity of International Organizations to Conclude Treaties and the Special Legal Aspects of the Treaties So Concluded* (The Hague, 1966).

5. See Oppenheim, *International Law*, I, 420–422. In its advisory opinion on "*Reparation for Injuries Suffered in the Service of the United Nations*" the International Court of Justice took a similar position, concluding that the U.N. is "a subject of international law and capable of possessing international rights and duties," primarily from its capacity to conclude treaties, *ICJ Reports* 1949, p. 179.

6. K'ung Meng: "In order to prove their [bourgeois] theory that international organizations are subjects of international law, some bourgeois jurists base their reactionary viewpoint on the right of the United Nations to conclude treaties . . . It is true that the right to conclude international treaties is a right enjoyed by subjects of international law, and, according to the Charter, the United Nations Organization and its principal organs certainly do have the right to conclude treaties. But the activities of the United Nations and its specialized agencies are strictly regulated by their respective member states, and the scope of the right to conclude treaties is also extremely limited. They do not appear in the capacity of subjects of international law, possessing rights and bearing obligations independently, but rather in the capacity of representative of all member states (genuine subjects of international law), because these treaties only create rights and obligations for member states," in "A Criticism of the Theories of Bourgeois International Law on the Subjects of International Law and the Recognition of States," *KCWTYC*, 2:50–51 (1960).

7. *E.g.*, see this statement by a Soviet scholar of international law: "In modern bourgeois legal writings, a number of scholars (Jessup, Lauterpacht, Scelle, etc.)

But, according to Triska and Slusser,[8] the prevalent Soviet view now is that international organizations are subjects of international law, albeit limited ones. There does not appear to be any writer in Communist China who advocates this new Soviet doctrine.

The practice of Communist China does not seem to exclude international organizations from concluding treaties. The agreement establishing the International Organization for the Cooperation of Railroads, in which Communist China participated, expressly provides that the legal status of its employees in member states will be jointly decided by the organization together with governmental organs of the states concerned.[9] Presumably, this could be decided through the conclusion of a separate treaty. Also, on July 27, 1953, the Communist Chinese Commander of the Chinese People's Volunteers in Korea, together with the Supreme Commander of the Korean People's Army, signed the Korean Armistice Agreement with the United Nations Command in Korea.[10] The agree-

favour the extension of the range of subjects of international law to include international organizations . . . But, this contradicts the very essence of international law as inter-State law . . . No international organizations . . . can be subjects of international law," Yevgenyev, "The Subject of International Law," in Kozhevnikov, ed., *International Law* (Moscow, 1961), p. 89. This is a translation of the 1957 Russian version with the addition of some post-1957 events; the Russian version was translated into the Chinese in 1959. But this is not the current Soviet view of the international personality of international organizations.

8. *The Theory, Law, and Policy of Soviet Treaties* (Stanford, 1962), pp. 49–52.

9. Art. 7, par. 1, of Rules Governing the Committee of International Organization for the Cooperation of Railroads (Peking, June, 1962), *TYC*, 11:141, 155.

10. It is not clear whether the signature "Commander-in-Chief, United Nations Command," can attribute the agreement to the U.N. The General Assembly, however, adopted Resolution 711 (VII), noting "with approval the Armistice Agreement concluded in Korea on 27 July 1953," U.N. General Assembly, *Official Records*, 7th sess., supp. no. 20B (A/2361/add. 2, 1953), p. 1. On another occasion, the U.N. presented a claim to the opposing side based on the Armistice Agreement in a manner normally employed only by parties to the agreement: Resolution 906 (IX) adopted Dec. 10, 1954, in which the General Assembly

"1. *Declares* that the detention and imprisonment of the eleven American airmen, members of the United Nations Command, referred to in document A/2830, and the detention of all other captured personnel of the United Nations Command desiring repatriation is a violation of the Korean Armistice Agreement;

"2. *Condemns*, as contrary to the Korean Armistice Agreement, the trial and conviction of prisoners of war illegally detained after 25 September 1953;

"3. *Requests* the Secretary-General, in the name of the United Nations, to seek the release, in accordance with the Korean Armistice Agreement, of these eleven United Nations Command personnel, and all other captured personnel of the United Nations Command still detained," U.N. General Assembly, *Official Records*, 9th sess., supp. no. 21, (A/2890, 1955), p. 56. But, on the other hand, the Armistice Agreement was not filed and recorded ex officio by the U.N. Secretariat. This process is required (by the Regulations on Registration) of every agreement to which the U.N. is party. Furthermore, it is interesting that the United States did ment was later included in the appendix to the Communist Chinese official Compilation of Treaties (*TYC*) under the heading "military affairs." The appendix of the Compilation usually contains only semiofficial agreements.[11] But in the cumulative "state index" of the official Compilation, the agreement was listed under the heading "United Nations."[12]

The capacity of a dependent state (a vassal state or a protected state) to conclude treaties depends upon the particular relationship between the dominant state (protecting state or suzerain state) and the dependent state concerned. The views of Western jurists are in conflict as to the validity of treaties concluded in defiance of a dependent state's restricted capacity. Admitting that this is "a question which does not admit of a clear answer," Lord Arnold McNair nevertheless concludes that "in most cases, such a treaty would be void."[13] The Communist Chinese position is generally the same as the Western view presented by McNair.

In the correspondence between Premier Chou En-lai and Prime Minister Nehru concerning the Sino-Indian boundary question, Communist China insisted that no agreement between China's Tibetan local authorities and representatives of Great Britain could bind China because China had neither ratified nor recognized any such agreement.[14] Communist China was arguing in effect that the Sino-Indian boundary had never been formally fixed by a treaty.

---

not register the agreement with the U.N.; such registration is required under art. 102 of the U.N. Charter if the United States considers itself party to the Armistice Agreement; see Chiu, *The Capacity of International Organizations to Conclude Treaties*, pp. 89–90; Seyersted, *United Nations Forces in the Law of Peace and War* (Leyden, 1966), pp. 103–108.

11. *TYC*, 2:382.

12. *TYC*, 11:224. During the armistice negotiations and the sebsequent Political Conference at Geneva (1951–1954), Communist China treated or spoke of the U.N. as party to the conflict and a "belligerent," see Goodrich, "Korea: Collective Measures against Aggression," *International Conciliation*, 494:182 (October 1953). Recent Communist Chinese statements specifically designate the U.N. as "a belligerent party and an aggressor" in the Korean conflict, *e.g.*, Communist Chinese Foreign Ministry statement (Sept. 28, 1965), *People's Daily*, Sept. 29, 1965, p. 4.

13. McNair, *Law of Treaties* (London, 1961), p. 42.

14. See *The Sino-Indian Boundary Question* (Peking, 1962), pp. 57–58. Wei Liang writes: "With respect to the exchange of letters carried out secretly between Britain and Tibetan local authorities in order to fabricate the so-called McMahon line, it was completely illegal. We have already stated . . . that treaties are agreements between states which should be formally signed by the plenipotentiary representative of the state concerned. Tibet, however, is only a part of the Chinese territory and the representative of the Tibetan local authorities could not represent the Chinese government. Therefore, the letters he exchanged secretly with the British representatives absolutely could not constitute an agreement between the Chinese and British governments," "The So-called McMahon Line," p. 49.

Nevertheless, it would still be incorrect to say that the Communist Chinese position on the treaty-making capacity of dependent entities is exactly the same as the view held by most Western jurists. Although Communist China categorically denies that Tibet possesses any independent treaty-making capacity,[15] and terms any assertion that Tibet is (or was) an independent state as a conspiracy of British imperialism, its attitude toward Outer Mongolia has been different. Outer Mongolia was a Chinese dependency from 1689 until 1945, when the Republic of China was forced to recognize its independence. In 1921 Soviet Russian forces in Siberia invaded Outer Mongolia on the pretext that they were pursuing White Russian "Guardists." Shortly thereafter, the "Mongolian People's Revolutionary Party" set up a "Mongolian People's Revolutionary Government." In May 1924, Soviet Russia signed an agreement with China which stated (Article 5) that "the Government of the Soviet Socialist Republics recognizes that Outer Mongolia is an integral part of the Republic of China and respects China's sovereignty therein."[16]

In July 1924, with Soviet forces still stationed in Outer Mongolia, Outer Mongolia declared its independence. This action, of course, was not recognized by the Chinese Central government. On March 12, 1936, Soviet Russia announced the conclusion of a "Mutual Assistance Pact" with Outer Mongolia. The Chinese government sent a strong note to the Soviet government on April 5, 1936, protesting conclusion of the agreement. Since the so-called independence of Outer Mongolia was not recognized at that time by the Chinese central government, the Soviet-Mongolian treaty should have been deemed void, consistent with the principles relied upon by Communist China in the Sino-Indian boundary dispute. However, although the Communist Chinese government has not made an official statement of the present Chinese attitude concerning the 1936 treaty, Communist Chinese writers appear to assume the validity of any treaties concluded between Soviet Russia and Outer Mongolia.[17] This appears to be because of the PRC's recognition of the People's Republic of Mongolia as an independent state in 1950, as part of the series of agreements then concluded with the Soviet Union.

15. Some Western jurists argue that Tibet was independent before 1950 and thereby had capacity to conclude treaties; *e.g.*, see *Tibet and the Chinese People's Republic* (Geneva, 1960), done by the International Commission of Jurists – which does not include Communist or Nationalist Chinese scholars.

16. *LNTS*, 37:175, 178 (effective May 31, 1924).

17. See P'an Lang, *Meng-ku jen-min kung-ho-kuo* (The Mongolian People's Republic) (Peking, 1950), pp. 58–59; Shu Yuan's book of the same title (Peking, 1961), p. 73. But it was reported in a *Pravda* editorial (Sept. 2, 1964) that Mao

Western scholars of international law generally agree that when an insurgent group fulfills several conditions it can be recognized as a belligerent and thereafter is a limited subject of international law.[18] In Communist Chinese legal literature, there has been no discussion of the status of insurgents who have been recognized as belligerents. But K'ung Meng does argue that "subjects" of international law should include nations in the course of struggling for independence or establishing their own states.[19]

Communist Chinese practice seems to indicate that revolutionary groups have the capacity to conclude treaties. Thus, before the establishment of the Central People's Government on October 1, 1949, the Northeastern (Manchurian) People's Government concluded a trade agreement with the Soviet Union in July 1949.[20] Moreover, before French recognition of the independence of Algeria in 1962, Communist China and the Provisional Government of Algeria issued several joint communiqués;[21]

Tse-tung did raise with Khrushchev the question of returning Outer Mongolia during the latter's visit to Communist China in 1954; see *International Affairs*, 10:83 (Moscow, 1964). In 1950 a Communist Chinese writer differentiated the Tibetan and Mongolian situations as follows: "Mongolian independence is just because it is the revolutionary action of the Mongolian people to resist internal and external oppression, to attain national independence and democratic rights . . . But the present so-called Tibetan 'independence' is the action of the present Tibetan ruling class--the Dalai Lama and his ruling clique--at the instigation of the American and British imperialists, to separate Tibet from the Chinese people and to lean to the imperialist camp . . . Therefore, Tibetan 'independence' and Mongolian independence are essentially different," Hai Fu, *Wei shih-ma i-pien-tao* (Why lean to one side)? (Peking, 1951), p. 87. Recently Communist China accused the Mongolian government of selling Mongolia's "sovereignty and independence" to the Soviet Union; *e.g.*, see Communist China's protest to Mongolia (Aug. 18, 1967), "Chinese Foreign Ministry Refutes Mongolian Government's Anti-China Statement," *PR*, 10:26, no. 35 (Aug. 25, 1967).

18. See Oppenheim, I, 140–141; Oppenheim, *International Law*, II, 7th ed. (London, 1952), 249–250.

19. "A Criticism of the Theories of Bourgeois International Law," p. 49.

20. Slusser and Triska, *A Calendar of Soviet Treaties, 1917–1957* (Stanford, 1959), pp. 261–262. The agreement was not included in the official Compilation of Treaties. The head of the Northeastern People's Government was later purged for separatist tendencies; Chin Szu-k'ai, *Communist China's Relations with the Soviet Union, 1949–1957* (Hong Kong, 1961), p. 4. In August 1948 the Chinese Communists reached an "understanding" with UNICEF on the latter's operation in Communist-held Chinese territory. The understanding was concluded by exchange of letters (dated Aug. 28 and 30) between Dr. Leo Eloesser, UNICEF rep. in North China, and the China Liberated Areas Relief Association; see "UNICEF in North China Violates Shihchiachuang Understanding," *NCNA*, Daily News Release no. 244 (Jan. 1, 1950), pp. 4–6. For details, see Lee, "Treaty Relations of the People's Republic of China: A Study of Compliance," *U. Penn. L. R.*, 116:301–308 (1967), and his *China and International Agreements* (Leyden, 1969), pp. 106–115.

21. The joint communiqués were issued Dec. 20, 1958, May 19, 1960, and Oct. 5, 1960; *TYC*, 7:9, 9:18, 9:20.

those documents were formally included in that part of the official Compilation of Treaties (*TYC*) which contains joint communiqués issued with other states. This was in spite of the fact that at the time of issuance of those joint communiqués the Provisional Government of Algeria was in fact a government-in-exile which did not have effective control over Algeria.

Nomenclature and Form of Treaties

International law does not prescribe any particular names for treaties. They can be termed not only treaties, but also agreements, acts, conventions, declarations, protocols, exchanges of notes, and so forth. The international juridical effect of a treaty is not dependent upon the name given to the instrument. Communist China seems to accept and follow these principles in its treaty practice.[22] The official Compilation of Treaties,[23] put together by the Communist Chinese Foreign Ministry, contains international documents with various titles, such as treaties, agreements, protocols, exchange of notes, joint communiqués, and declarations.

Despite the lack of legal significance of the names of treaties, international treaty practice indicates that certain types of treaties do customarily go under certain names.[24] With respect to this point, Communist Chinese writer Wang Yao-t'ien explains six different names used for treaties.

> (1) Treaty – This name is used to designate the most important of international documents, regulating the political, economic or other relations between contracting states, such as a treaty of alliance and mutual assistance or a treaty of commerce and navigation.
> (2) Agreement – A treaty regulating special or provisional problems of the contracting states is called an "agreement," such as a trade agreement or a payment agreement.
> (3) Convention – An agreement regulating special problems among several states is called a "convention" [Kung-yüeh] such as a postal convention or a telecommunication convention. A bilateral agreement

22. See Wang Yao-t'ien, *International Trade Treaties*, p. 12.
23. Reviewed by Chiu in *AJIL*, 61:1095 (1967). *TYC* does not include all treaties or agreements concluded by Communist China; for a more complete listing, see Johnston and Chiu, *Agreements of the People's Republic of China, 1949–1967* (Cambridge, Mass., 1968).
24. *E.g.*, see Starke, *Introduction to International Law* (London, 1967), pp. 340–343.

of this type is generally translated into Chinese as "chuan-yüeh" [Convention], such as a consular convention or a boundary convention.

(4) Declaration – This is an international document which generally provides only for general principles of international relations and international law. Sometimes it also provides for specific obligations, such as the 1856 Paris Declaration concerning the law of sea warfare or the Cairo Declaration of December 1, 1943.

(5) Protocol – This is an international document containing an agreement on individual problems. Sometimes it amends, interprets or supplements certain provisions of a treaty, such as the general conditions for the delivery of goods concluded by foreign trade ministries of socialist states or the Soviet-Japanese protocol on reciprocal application of most-favored-nation treatment concluded on October 19, 1956.

(6) Exchange of notes – These are notes exchanged between two states to define certain matters already agreed upon by them.[25]

While some Western jurists suggest that a treaty can be concluded in oral form,[26] Communist China's practice suggests that it prefers treaties to be in written form.[27] This conforms to recent international treaty practice which prefers treaties in written form.[28]

25. Wang Yao-t'ien, *International Trade Treaties*, p. 12.

26. *E.g.*, see Oppenheim, *International Law*, I, 898.

27. Cf. Wang Yao-t'ien *International Trade Treaties*, p. 13: "written treaties are the typical form of modern international treaties." But the question whether Communist China ever accepts oral agreements as treaties is unclear; *e.g.*, no written document was reported signed by Communist China and Hong Kong authorities re the agreement between them to release two Hong Kong policemen who strayed into Communist China in September 1967. See "Crisis in Hong Kong Is Eased by Accord," *New York Times*, Dec. 1, 1967, pp. 1, 16.

But when on Dec. 3, 1962, the Indian government requested the Chinese government to close its consulates in India, Communist China protested to the Indian government the "unilateral tearing up of an agreement between the two governments," see "Note Given by the Ministry of Foreign Affairs, Peking, to the Embassy of India in China, 8 December 1962," in *Notes, Memoranda and Letters Exchanged between the Governments of India and China (October 1962–January 1963), White Paper,* no. VIII (New Delhi, 1963), pp. 123–124. The Indian government denied that there was any such agreement in existence, though it recognized that "as for the Chinese Consulate-General in Bombay, the Government of India had agreed in 1952 to its being established on an *ad hoc* basis at the time of the re-designation of the Indian Mission in Lhasa a Consulate-General," "Note Given by the Ministry of External Affairs, New Delhi, to the Embassy of China in India, 21 March 1963," *ibid.*,*(January 1963–July 1963), White Paper*, no. IX, pp. 175–176.

Possibly Communist China regarded the 1952 arrangement as a binding oral agreement between China and India; see Charng-ven Chen, "Communist China's Attitude Toward Consular Relations" (unpub. seminar paper, Harvard, 1970), pp. 97–100, 203–204.

28. Art. 2(a) of the 1969 Convention on the Law of Treaties provides that "treaty" means an agreement concluded between states in written form. But art. 3

The form in which treaties are concluded does not in any way affect their legally binding force. Thus a treaty may be embodied in a single instrument or in two or more related instruments. Moreover, a treaty may be concluded in the form of an agreement between heads of states or governments, states, ministers of states, departmental heads of governments, and others. Communist China's practice follows these principles.

Despite agreement on the above-stated principles, however, the United States and Communist China recently have subjected these principles to conflicting interpretations in at least two important instances. The first concerns the legally binding force of certain declarations made during the Second World War. On December 1, 1943, the United States, the United Kingdom, and China issued the Cairo Declaration, which stated that "all the territories Japan has stolen from the Chinese, such as Manchuria, Formosa, and the Pescadores, shall be restored to the Republic of China."[29] On July 26, 1945, the three states issued the Potsdam Proclamation, which promised, inter alia, that "the terms of the Cairo Declaration shall be carried out."[30] The Soviet Union later adhered to the Proclamation; and on September 2, 1945, Japan accepted it in her Instrument of Surrender.[31] On October 25, 1945, Taiwan (Formosa) was restored to China – which soon declared Taiwan to be its thirty-fifth province. At that time, no objection was raised against the retrocession of this former Chinese territory to China.

The United States position on the Cairo Declaration and Potsdam Proclamation is inconsistent. On January 5, 1950, President Truman stated that "in keeping with these declarations, Formosa was surrendered to Generalissimo Chiang Kai-shek, and for the past 4 years, the United States and the other Allied Powers have accepted the exercise of Chinese authority over the Island."[32] On the same day the Secretary of State

---

provides that "the fact that the present Convention does not apply to . . . international agreements not in written form, shall not affect: (a) the legal force of such agreements; (b) the application to them of any of the rules set forth in the present Convention to which they would be subject under international law independently of the Convention," see *ILM*, 8:681–682, no. 4 (July 1969).

29. "Conference of President Roosevelt, Generalissimo Chiang Kai-shek and Prime Minister Churchill in North Africa," *DSB*, 9:393, no. 232 (Dec. 4, 1943); *Foreign Relations of the United States, Diplomatic Papers: The Conferences at Cairo and Teheran 1943* (Washington, 1961), pp. 448–449.

30. "Proclamation Defining Terms for Japanese Surrender," *DSB*, 13:137, no. 318 (July 29, 1945); *Foreign Relations of the United States, Diplomatic Papers: The Conference of Berlin (The Potsdam Conference), 1945* (Washington, 1960), pp. 1474–1476.

31. *EAS*, no. 493, 59 Stat. 1733 (1945).

32. "United States Policy Toward Formosa," *DSB*, 22:79, no. 550 (January 16, 1950).

elaborated: The Cairo Declaration "was incorporated in the declaration at Potsdam and that declaration . . . was conveyed to the Japanese as one of the terms of their surrender and was accepted by them, and the surrender was made on that basis. Shortly after that, the Island of Formosa was turned over to the Chinese in accordance with the declarations made and with conditions of the surrender. The Chinese have administered Formosa for 4 years. Neither the United States nor any other ally ever questioned that authority and that occupation. *When Formosa was made a province of China nobody raised any lawyers' doubts about that. That was regarded as in accordance with the commitments*"[33] (emphasis added). After outbreak of the Korean War the United States suddenly changed position. On June 27, 1950, President Truman said that "determination of the future status of Formosa must await the restoration of security in the Pacific, a peace settlement with Japan, or consideration by the United Nations."[34] No legal ground was set forth to explain how the United States position could be reconciled with the commitments embodied in the Cairo Declaration and Potsdam Proclamation.

On December 1, 1954, Secretary of State Dulles said in a press conference: "technical sovereignty over Formosa and the Pescadores has never been settled . . . the future title is not determined by the Japanese peace treaty, nor is it determined by the peace treaty which was concluded between the Republic of China and Japan." On December 10, Dulles addressed a note to Minister of Foreign Affairs of the Republic of China George K. C. Yeh stating in part that "the Republic of China effectively controls both the territory [Formosa and Pescadores] described in Article VI of the Treaty of Mutual Defense between the Republic of China and the United States of America signed on December 2, 1954, at Washington and other territory."[35]

33. *Ibid.*, p. 80.

34. "U. S. Air and Sea Forces Ordered into Supporting Action," *DSB*, 23:5, no. 547 (July 3, 1950).

35. Quoted from Whiteman, *Digest of International Law*, 3:564 (Washington, 1964).

British Foreign Secretary Sir Anthony Eden stated on Feb. 4, 1955, that the Cairo Declaration "was a statement of intention that Formosa should be retroceded to China after the war," House of Commons, *Debates*, 5th ser., 536:159 (London, 1955). For a study of the British attitude toward the Cairo Declaration and the legal status of Formosa see Jain, "The Legal Status of Formosa," *AJIL*, 57:27–32 (1963).

The treaty character of the Cairo and Potsdam Declarations may be doubtful, but the terms of both were included in the Japanese Instrument of Surrender. The latter's treaty character seems beyond doubt in that, as in the case of other treaties concluded by the U.S., it was printed in *United States Statutes at Large*, 59 Stat. 1733 (1945).

The official attitude of Communist China was expressed in a cablegram sent to the United Nations on August 24, 1950, which read in part: "Taiwan is an integral part of China. This is . . . stipulated in the Cairo Declaration of 1943 and the Potsdam Communiqué of 1945 as binding international agreements which the United States Government has pledged itself to respect and observe."[36] Supporting his government's position, Communist Chinese writer Shao Chin-fu maintains, in an article entitled "The Absurd Theory of 'Two Chinas' and Principles of International Law," that the Cairo Declaration and Potsdam Proclamation are not merely broad statements of policy but are, instead, binding legal commitments by the participating heads of states to perform certain acts, including the restoration of Taiwan to China upon the defeat of Japan.[37] To support his argument, he cites a passage in the seventh edition of Lauterpacht's Oppenheim: "Official statements in the form of Reports of Conferences signed by the Heads of States or Governments and embodying agreements reached therein may, in proportion as these agreements incorporate definite rules of conduct, be regarded as legally binding upon the states in question."[38]

It may be interesting to know that the Communist Chinese position is also supported by the fact that since 1945 Taiwan has been administered by the Nationalist Chinese government, which has consistently maintained that the Cairo Declaration and the Potsdam Proclamation are legally binding. On February 8, 1955, President Chiang Kai-shek said:

> I recall that in 1943, the late American President Roosevelt and the present British Prime Minister Churchill and I held a conference at Cairo to discuss problems relating to the prosecution of war against Japan and its aftermath. In the communiqué issued at the conclusion of the conference, we announced that all the territories "stolen" by Japan from China, including the Northeast provinces and Taiwan and Penghu, should be returned to the Republic of China. This announce-

36. *WCC*, 1:134; U.N. Doc. S/1715 (1950).

37. *KCWTYC*, 2:13–14 (1959); English trans. in *Oppose the New U.S. Plots to Create "Two Chinas"* (Peking, 1962), p. 92.

38. *International Law*, I, 7th ed. (London, 1948), 788. Another Communist Chinese writer, Ch'en T'i-ch'iang, cited this paragraph: "The Cairo Declaration doubtless belongs to this category of international documents, that is to say, it is an international document 'legally binding upon the States in question.' Furthermore, the Potsdam Proclamation issued by China, the United States, and Great Britain on July 26, 1945 to urge the surrender of Japan reaffirmed the obligations of the Cairo Declaration. It was provided in the Potsdam Proclamation that 'the terms of the Cairo Declaration shall be carried out.' The phrase 'shall be carried out' conclusively proves that the Cairo Declaration is a document creating interna-

ment was recognized by the Potsdam Declaration and accepted by Japan at the time of her surrender. Its validity is thus based on a number of agreements and should not be questioned.

Therefore, when Japan surrendered, the Government of the Republic of China repossessed Taiwan and Penghu and constituted them as Taiwan Province. Since that time, Taiwan and Penghu have regained their status as an integral unit of the territory of the Republic of China. In the San Francisco Peace Treaty and the Sino-Japanese Peace Treaty, Japan renounced her sovereignty over Taiwan and Penghu, thereby completing the process of restoring these areas to our country . . .

Some people deny the validity of the Cairo Declaration in order to justify their distorted view about the status of Taiwan. If one could deny the validity of the Cairo Declaration, what about the Potsdam Declaration and all the international treaties and agreements concluded since the termination of the Second World War? Could their validity be also denied? If the democracies repudiate the Cairo Declaration, which they signed themselves, how, either now or in the future, can they criticize the Communist aggressive bloc for tearing up treaties and agreements? Those who play fast and loose about the status of Taiwan do so against their own conscience. In their eagerness to appease the aggressors they have their eyes fixed on a transient state of affairs. They do not realize how gravely they jeopardize world security by arguing from bad law and false policies.[39]

A second dispute arose from the "agreed announcement" issued by representatives of Communist China and the United States in 1955 on

---

tional obligations, and not merely a statement of the intentions of the signatories. Upon the participation of the Soviet Union in the Potsdam Proclamation and the acceptance by Japan of the terms of the Proclamation by its unconditional surrender on September 2, 1945, the obligations of the Cairo Declaration became obligations binding upon the Four Great Powers as well as Japan.

"From a doctrinal view of international law, it is impossible to question the binding force of the Cairo Declaration as an international treaty," in "Sovereignty of Taiwan Belongs to China," *People's Daily*, Feb. 8, 1955, p. 4; English trans. entitled "Taiwan – A Chinese Territory," *Law in the Service of Peace* (Brussels), 5:39–40 (1956).

39. "Review of International Situation," in *President Chiang Kai-shek: Selected Speeches and Messages in 1955* (Taipei, 1956), pp. 22–23.

At the proposal of the United States, the 1951 San Francisco Peace Treaty with Japan (effective Apr. 28, 1952) provided (art. 2) that Japan renounce all "right, title and claim" to Taiwan but did not mention the return of Taiwan to China; *UNTS*, 136:35. This renunciation was confirmed by art. 2, Republic of China–Japan Peace Treaty of Apr. 28, 1952, *UNTS*, 138:3 (effective Aug. 5, 1952). Neither treaty expressly provided for the transfer of sovereignty over Taiwan to China or any other authority. Western countries have maintained that because of that lack the status of Taiwan is "undetermined" and the island is not legally part of China, despite the Republic of China's exercise of control over the island since Oct. 25, 1945. Ely Maurer, then Assistant Legal Adviser for Far Eastern Affairs, the question of the return of each other's civilians.[40] The text of the "agreed announcement" took the form of statements issued separately by the ambassadors of the two countries. The first part is the statement of the United States Ambassador, informing the Communist Chinese Ambassador of the measures to be taken by the United States; the second part is the statement of the Communist Chinese Ambassador, informing the United States Ambassador of the measures to be taken by Communist China. Communist China regards the "agreed announcement" as an international treaty, and it was accordingly included in Volume 4 of its official Compilation of Treaties.[41] But the legal status of the "agreed announcement" in the United States is vague. It is neither printed in the official *United States Treaties and Other International Acts* series nor listed in the *Treaties in Force* published by the Department of State.[42] The United States, apparently, does not consider it to be either a "treaty" or an "executive agreement." Although the United States government did take certain steps to implement the "agreed announcement," Communist China was not satisfied and charged that the United States had violated it in a number of cases.[43] Thus when the United States in 1960 presented to the Communist Chinese Ambassador in Warsaw a draft agreement, in the same form, concerning an exchange of correspondents,

---

U.S. Department of State, wrote in "Legal Problems Regarding Formosa and the Offshore Islands" that "the situation is . . . one where the Allied Powers still have to come to some agreement or treaty with respect to the status of Formosa," *DSB*, 39:1009, no. 1017 (Dec. 22, 1958).

Some Western scholars have taken a different view toward the Anglo-American position on the status of Taiwan, *e.g.*, D. P. O'Connell, of Australia: after Japanese renunciation of the island it is "doubtful . . . whether there is any international doctrine opposed to the conclusion that China appropriated the *terra derelicta* of Formosa by converting belligerent occupation into definitive sovereignty," in "The Status of Formosa and the Chinese Recognition Problem," *AJIL*, 50:405, 415 (1956). Frank Morello, of the U.S., has also argued that the Republic of China has acquired sovereign right over Formosa in accordance with the international law principle of prescription; *The International Legal Status of Formosa* (The Hague, 1966), p. 98.

40. "U.S., Red China Announce Measures for Return of Civilians," *DSB*, 33:456, no. 847 (Sept. 19, 1955).

41. *TYC*, 4:1 (1955).

42. It was referred to by an American court as a "written agreement" between the U.S. and Communist China; *in re* Eng's Estate, District Court of Appeal, 2nd Dist. (June 24, 1964), 228 Calif. App. 2d 160; *California Report* (1964), p. 254.

43. *E.g.*, see Communist Chinese Foreign Ministry statement of Mar. 11, 1956, *WCC*, 4:52. The U.S. also charged that Communist China failed to implement the "agreed announcement"; *e.g.*, see statement by U.S. Department of State of Dec. 16, 1955; "Continued Detention of U.S. Civilians by Communist China," *DSB*, 33:1049, no. 861 (Dec. 26, 1955).

Communist China, in a statement issued on September 13, 1960, rejected the draft on these grounds:

> We remember that the agreement of the two sides on the return of civilians to their countries also took this form. But the fact that the U.S. side has so far failed to seriously implement the agreement shows that this form does not have enough binding force on the U.S. side. To prevent the U.S. side from again violating [the] agreement, the Chinese side resolutely maintains that all agreements between the two sides must take the form of joint announcements of both sides, and no longer take that of statements issued by two sides separately.[44]

Classification of Treaties

In the official Compilation of Treaties of the People's Republic of China, treaties are classified into fourteen categories; five of these categories are further classified into subcategories:

1. Political
    (1) Friendship
    (2) Joint announcement, communiqué, or declaration
    (3) Others
2. Legal
    (1) Consular relations
    (2) Nationality
3. Boundary
4. Boundary problems (use of boundary river, etc.)
5. Economic
    (1) Commerce and navigation
    (2) Economic aid, loan, and technical cooperation
    (3) Trade and payment
    (4) General conditions for delivery of goods
    (5) Registration of trademark
    (6) Others
6. Cultural
    (1) Cultural cooperation
    (2) Broadcasting and television cooperation

44. Statement of Sept. 13, 1960, issued by Communist Chinese Foreign Ministry, *WCC*, 7:244, 246; English trans. in "Refuting U.S. State Department: Chinese Statement on the Question of Exchanging Correspondents Between China and the U.S.," *PR*, 3:30, no. 37 (Sept. 14, 1960).

(3) Exchange of students
(4) Others

7. Science and technology
8. Agriculture and forest
9. Fishery
10. Health and sanitation
11. Post and telecommunication
12. Communication and transportation
    (1) Railways
    (2) Air transportation
    (3) Water transportation
    (4) Highway
13. Law of war
14. Military

In addition to the above-stated classifications of treaties, it should be noted that some agreements concluded by Communist Chinese people's organizations (nongovernmental organs) are of considerable importance. They may be classified as "semiofficial" agreements. In the official Compilation, such agreements appear in the appendix. They are usually, but not necessarily, concluded between Communist Chinese people's organizations and similar organizations in states which do not maintain diplomatic relations with Communist China. Sometimes they in fact, if not in law, involve official functions between Communist China and other states.[45] For instance, several agreements concluded between the Chinese Fishery Association and the (Japanese) Sino-Japanese Fishery Association are of this character, for they cover such important matters as salvage of fishing boats at sea, conservation of fishing resources, demarcation of fishing zones on the high seas, and allocation of the number of fishing boats each side may send into the zones.[46] Table 1 lists states whose semiofficial agreements with Communist China appeared in the 1949–1964 Compilation.

Furthermore, the local authorities of Kwangtung Province occasionally have concluded agreements with the Portuguese authorities in Macao or the British authorities in Hong Kong. I propose to call these "local" agreements. They usually deal with local matters, such as the supply of

45. For a brief discussion of Communist China's semiofficial trade agreements, see Hsiao, "Communist China's Trade Treaties and Agreements," *Vanderbilt L.R.*, 21:631–632 (1968).
46. *TYC*, 4:265, 276, 280; 5:411, 413; 6:330, 331, 333.

Table 1. States and organizations whose semiofficial agreements with Communist China appeared in the 1949-1964 Compilation

State	Number of semiofficial agreements concluded	Observation
Austria	2	–
Finland	2[a]	Concluded between several Chinese state export trading companies and the Finnish Ministry of Industry and Commerce
France	3	All concluded before French recognition of Communist China in 1964
West Germany	4	–
Japan	24	–
Singapore and Malaya	1	–
United Kingdom	1[a]	–
United Nations	1	Korean Armistice Agreements, July 27, 1953; signed by Commander-in-chief of Chinese People's Volunteers

[a]State maintaining diplomatic relations with Communist China.

water[47] and the settlement of local disputes.[48] Communist China does not include these agreements in the official Compilation, though the *NCNA* occasionally reports their conclusion in the *People's Daily*[49] or in radio broadcasts.

47. *E.g.*, a Reuters news release from Hong Kong of Oct. 1, 1967, disclosed that in 1964 Communist China concluded an agreement with Hong Kong British authorities re the 15-billion-gallon annual water supply to Hong Kong; "China Resumes Supply of Water to Hong Kong," *New York Times*, Oct. 1, 1967, p. 6.

48. On Jan. 29, 1967, Portuguese authorities in Macao signed a document admitting responsibility for the killing of several Chinese residents in Macao on Nov. 15 and Dec. 3, 1966. They also signed a protocol accepting the 4-point demand for settlement of the incident and other matters put forth by the Director of the Foreign Affairs Bureau of the Kwangtung Provincial People's Council on Dec. 9, 1966; "Victory for the Patriotic Anti-Imperialist Struggle by Compatriots in Macao," *PR*, 10:31, no. 6 (Feb. 3, 1967).

49. *E.g.*, the above-mentioned 4-point demand was published in *People's Daily*, Dec. 10, 1966, p. 2.

As to the opinions of Communist Chinese writers concerning the problem of classification of treaties, although various writers refer to "equal" and "unequal" treaties, and "commercial treaties" and "trade agreements," only one offers a general discussion of this problem. According to Wang Yao-t'ien, there is no generally recognized criterion for classifying international treaties and agreements.[50] He points out that treaties can be classified as bilateral, trilateral, and multilateral, corresponding to the number of contracting parties. He also observes that some scholars look to the importance of the contents of treaties for a classificatory criterion, dividing them into lawmaking treaties and general administrative agreements. Duration is a third criterion Wang sees; treaties can be for an indefinite or definite period, and the latter category can be subdivided into short-term and long-term treaties.

Finally, Wang discusses at some length the classification of treaties according to their contents. He finds four types:

(1) Political treaties and agreements, such as treaties of alliance, mutual nonaggression, and peace

(2) Economic treaties and agreements, such as treaties of commerce and navigation and agreements of trade, economic cooperation, payment, clearance, barter, tariff, transportation, communication and fisheries

(3) Social welfare treaties and agreements, such as narcotics agreements and sanitary conventions

(4) Cultural and educational treaties and agreements, such as cultural agreements and scientific and technical cooperation agreements

Wang notes that this classification is not inflexible. In many instances, economic clauses are included in political treaties and agreements, and economic treaties and agreements may include clauses dealing with social welfare and cultural matters.

Another Communist Chinese writer, Wang Yih-wang, discusses briefly the difference between "commercial treaties" and "trade agreements."[51] In his view a trade agreement usually contains "the total amount of export and import of both states, commodities in trade, method of delivery, method of clearing the difference between export and import, tariff, etc." The duration of a trade agreement is relatively short, usually from

50. *International Trade Treaties*, p. 11.

51. Wang Yih-wang, "What Are the Differences between Commercial Treaties and Trade Agreements," *Enlightenment Daily*, May 12, 1950, p. 3.

one to five years. The scope of a "commercial treaty," on the other hand, is relatively broad and includes the overall commercial relations between contracting parties. It deals more with matters of principle rather than with the concrete items provided for in a trade agreement and it usually covers a longer period of time. The titles, "commercial treaty" or "trade agreement," are not necessarily determinative of the nature of the document. According to Wang Yih-wang, "American imperialists have frequently used the title of a trade agreement to incorporate the contents of a . . . commercial treaty."[52]

52. See also Wang Yao-t'ien, *International Trade Treaties*, pp. 19–25, where the problem is discussed in greater detail.

III Conclusion and Entry into Force of Treaties

Subject only to the limitations imposed by international law, treaties, or its own domestic law, the government of a state is competent to conclude any treaty on behalf of that state. Thus, Communist Chinese writer Wang Yao-t'ien writes that "the right to conclude international treaties is one of the fundamental rights of all sovereign states."[1]

Communist Chinese practice and literature do not generally challenge this principle, but they do indicate that the competence of the governments of certain states, especially divided states, to conclude certain treaties is subject to some limitations. A few cases illustrating such limitations follow.

(1) On October 27, 1956, the governments of the Federal Republic of Germany and France concluded a treaty providing for the return of the Saar to Germany.[2] On January 2, 1957, an article in the *People's Daily* commented that "at the present time, when two German states exist, the question of the Saar can be reasonably solved only when representatives of the two German states participate in the settlement." The article also supported East Germany's protest against this settlement of the Saar problem.[3]

(2) On June 22, 1965, the government of the Republic of Korea signed a "basic treaty" with Japan to normalize their relations.[4] On June 23, 1965, North Korea made a statement declaring that the treaty is "completely null and void." On June 26, 1965, Communist China stated that it "firmly support[s] the just stand" of North Korea and that it "will never recognize the so-called 'ROK-Japan Basic Treaty.'"[5]

(3) If the government of a state is considered by Communist China as a "reactionary government," it is judged incompetent to conclude agreements that invite foreign assistance in coping with cases of domestic disturbance.[6] The appeal of Lebanon's President Chamoun for the dispatch

1. Wang Yao-t'ien, *Kuo-chi mao-yi t'iao-yüeh ho hsieh-ting* (International trade treaties and agreements) (Peking, 1958), p. 12.

2. *Journal official: lois et decrets*, Jan. 10, 1957 (Paris, 1957), p. 460.

3. Li Pan-to, "Saar Treaty--A Dirty Deal," *People's Daily*, Jan. 2, 1957, p. 6.

4. *ILM*, 4:924, no. 5 (September 1965).

5. *People's Daily*, June 27, 1965, p. 1; English trans. in "Chinese Government Statement: China Will Never Recognize 'ROK-Japan Basic Treaty,' June 26," *PR*, 8:8, no. 27 (July 2, 1965).

6. *E.g.*, Shih Sung *et al.*: "In accordance with modern international law, a change in the form of the government of a state (this change may be conducted in accordance with the legal process of the state, or it may be the result of mass revolution), or the overthrow of the dynasty of a state are both matters which concern only the people themselves of that state and the internal affairs of the state.

of American troops in 1958 was considered invalid.[7] On the other hand, a "people's government" is certainly competent to conclude an agreement with another "people's government" to suppress "reactionary rebellion" which has broken out in that country. Thus, in accordance with the 1955 Warsaw Treaty[8] the Kadar government's request for Soviet intervention in Hungary in 1956 was considered "valid."[9]

(4) When two governments are contesting for the control of a state, the government regarded by Communist China as "reactionary" or the "puppet of American imperialism" is thought of as not competent to conclude certain treaties on behalf of that state. Thus, during the Chinese civil war, the Central Committee of the Chinese Communist Party issued a statement on February 1, 1947, that it "will not either now nor in the future recognize any foreign loans, any treaties which disgrace the country and strip away its rights, and any of the above mentioned agreements and understandings established by the Kuomintang government after January 10, 1946."[10] Recently, Communist China has asserted that the government of Vietnam no longer represents the South Vietnamese people

---

Other states have absolutely no right to intervene and even if the reactionary government had beforehand concluded a treaty with imperialism [to intervene] . . . other states also cannot conduct intervention, because this [treaty concluded] by the reactionary government violates the will of the people of that state and if other states conduct intervention, it is also an act in violation of international law," in "An Initial Investigation into the Old Law Viewpoint in the Teaching of International Law," *CHYYC*, 4:15 (1958).

7. *E.g.*, see Chou Keng-sheng, "Don't Allow American and British Aggressors to Intervene in the Internal Affairs of Other States," *CFYC*, 4:3 (1958).

8. *UNTS*, 219:3.

9. *E.g.*, see "Behind the Hue and Cry over the 'Hungarian Issue'" (editorial), *People's Daily*, Nov. 14, 1956, p. 1; English trans., same title, in *PC*, 23:5 (Dec. 15, 1956). In addition to the Warsaw Treaty, Ch'en T'i-ch'iang argues, the 1947 Hungarian Peace Treaty also authorized the Hungarian government to invite Soviet intervention: "The action of Soviet force was based on treaties. According to the Warsaw Treaty, the Soviet Union has the right and duty to take such 'concerted action' with Hungary 'as may be necessary to reinforce their defensive strength, in order to defend the peaceful labor of their peoples, guarantee the inviolability of their frontiers and territories, and afford protection against possible aggression' (Article 5). This provision is to be immediately put into effect in case of 'threat or armed attack' against Hungary. Moreover, in accordance with Article 4 of the 1947 Peace Treaty with Hungary [*UNTS*, 41:135], Hungary has the duty to dissolve 'all organizations of a Fascist type on Hungarian territory, whether political, military or para-military, as well as other organizations conducting propaganda hostile to the United Nations,' and 'shall not permit in the future the existence and activities of organizations of that nature which have as their aim denial to the people of their democratic rights.' To invite Soviet assistance when Hungary's own forces were not sufficient to carry out this obligation of the Peace Treaty is perfectly consistent with the spirit of the Treaty," in "The Hungarian Incident and the Principle of Non-Intervention," *Enlightenment Daily*, Apr. 5, 1957, p. 1.

10. *DIA*, 1947–1948, p. 664.

and that the National Front for Liberation is the "sole" representative of South Vietnam.[11] Presumably, in the Communist Chinese view the government of the Republic of Vietnam can no longer conclude treaties on behalf of South Vietnam.

Whether a treaty concluded in violation of the constitutional requirements of a state is binding is a controversial question among Western scholars.[12] This question has not been discussed by any Communist Chinese writer.[13] However, in an article by Professor Chou Keng-sheng on the 1958 Lebanon case, one of the reasons invoked against the legality of the United States' intervention in Lebanon was that President Chamoun's request for intervention was illegal under Lebanese law:

> As pointed out by Khrushchev, Chairman of the Council of Ministers of the Soviet Union, the "appeal was made by the irresponsible ruler of Lebanon without the support of the people and in disregard of the will of the people . . . and therefore it does not have any constitutional validity."[14] As a matter of fact, the speaker of the Lebanese Parlia-

11. See "The National Front for Liberation Is the Sole Representative of South Vietnamese People" (editorial), *People's Daily*, June 18, 1965, p. 1; English trans., same title, in *PR*, 8:10, no. 26 (June 25, 1965). Communist China did not challenge the competence of the government of the state of Vietnam (renamed Republic of Vietnam in 1955) to participate in the 1954 and 1961–62 Geneva Conferences and the 1955 Bandung Conference. The reasons for its change of attitude were explained in this editorial:

"Some people ask, why cannot the Saigon regime which took part in the First Afro-Asian Conference ten years ago attend the second conference? The reason is very simple. A tremendous change has taken place in the situation in South Vietnam since 1955. In the past ten years, U.S. imperialism has torn to shreds the Geneva agreements, undermined the peaceful reunification of Vietnam and followed up its intervention in south Vietnam with a large-scale war of aggression there. The Saigon authorities have long been puppets of U.S. imperialism, its tools of aggression and its agents to be used against the Asian and African peoples. They were long ago repudiated by the South Vietnamese people . . .

"The facts are crystal clear. Today the genuine representative of the south Vietnamese people can be none other than the N.F.L. which was born out of their just, patriotic struggle against U.S. aggression . . . Today the liberated areas controlled by the N.F.L. constitute four-fifths of the territory of south Vietnam and contain ten million people." See also "Vientiane Authorities Cannot Represent Laotian Peoples," *PR*, 8:11, no. 26 (June 25, 1965).

12. See Blix, *Treaty-Making Power* (New York, 1959), pp. 370–391.

13. Wang Yao-t'ien: "The right to conclude international treaties is generally a privilege of the supreme authority of a state. The constitution of each state shall decide which organ should exercise this power," *International Trade Treaties*, pp. 12–13. He did not discuss the question of validity of a treaty concluded in violation of a state's constitution.

14. I am unable to locate this passage in Khrushchev's statements during this period. In his July 19, 1958, message to President Eisenhower, Khrushchev stated:

ment had called the United Nations requesting the United States to withdraw its forces; this was, on behalf of the Lebanese people, a repudiation of Chamoun's appeal.[15]

In practice, some treaties concluded by Communist China do refer to the constitutional requirements of the contracting parties. For instance, Article 6 of the 1961 Sino-Indonesian Treaty of Friendship provides that: "The present treaty shall be ratified by the Contracting Parties in accordance with their respective constitutional procedures."[16] Another example is the 1957 Sino-Syrian Accord on Revising Trade Agreement and Payment Agreement between the two countries, which provides in Article 4 that: "This Accord shall be ratified or approved in accordance with their respective legal processes."[17] These examples seem to suggest that for Communist China the constitutional requirements of a state are not irrelevant in concluding a treaty.

Mutual Consent

According to the traditional Western view, a treaty concluded as the result of duress or coercion exercised personally against the representatives of a contracting state is invalid. On the other hand, if the coercion is exercised against the state itself, the treaty is valid. In recent years, however, Western scholars have expressed the view that, since resort to war or to the illegal use of force is prohibited by the 1928 General Treaty for the Renunciation of War and the 1945 United Nations Charter, duress exercised against the state must be regarded as vitiating any treaty that results.[18]

"The Lebanese Parliament, and the Speaker of the Parliament have strongly protested against American armed intervention. Consequently, the 'invitation' sent by Chamoun has no constitutional validity," "N. S. Khrushchev's Messages to President Eisenhower and Prime Ministers De Gaulle and Nehru," *Soviet News*, 3882:69 (July 22, 1958).

15. Chou, "Don't Allow American and British Aggressors to Intervene," p. 3.

16. *TYC*, 10:7; English trans. in "Treaty of Friendship between the People's Republic of China and the Republic of Indonesia," *PR*, 4:11, no. 24 (June 16, 1961).

17. *TYC*, 6:161.

18. See Oppenheim, *International Law*, I, 8th ed. (London, 1955), 891–892. Art. 51 of the 1969 Convention on the Law of Treaties provides that "the expression of a State's consent to be bound by a treaty which has been procured by the coercion of its representative through acts or threats directed against him shall be without any legal effect," art. 52 that "a treaty is void if its conclusion has been procured by the threat or use of force in violation of the principles of international law embodied in the Charter of the United Nations"; *ILM*, 8:698, no. 4 (July 1969).

Communist Chinese writers do not seem to have discussed these problems in general; but they have condemned many treaties concluded under coercion as illegal and void. In this connection, Communist Chinese writer Hsin Wu severely criticizes American scholar Charles Hyde's view that "the validity of a transfer of rights of sovereignty as set forth in a treaty of cession does not appear to be affected by the motives which have impelled the grantor to surrender them."[19] According to Hsin Wu:

> Obviously, according to such an interpretation, it was legitimate for Japan to force the Manchu Government of China to cede Taiwan and Penghu through the unequal Treaty of Shimonoseki after the 1895 Sino-Japanese War. This is tantamount to saying that it is legal for a robber to take property by brandishing a dagger before the owner, threatening his life, and then forcing him to put his fingerprint on a document indicating his consent. Is that not absurd? No wonder bourgeois international law has sometimes been described as the law of bandits.[20]

In addition to coercion, Communist Chinese writers consider the "unequal" or "aggressive and enslaving" nature of a treaty a vitiating condition. This is discussed in detail in Chapter IV.

No Communist Chinese writers appear to comment on the problem of physical coercion against the representatives of a contracting state. But, in view of their severe condemnation of coercion or other undue influence exercised against the contracting state itself, a fortiori one can safely conclude that they also consider that such acts vitiate a treaty.

Western jurists generally agree that a treaty is not binding if the consent of the contracting states was given in error or under a mistake produced by fraud on the part of a contracting state.[21] The opinion of Soviet jurists appears to be the same.[22] No Communist Chinese writers have ever commented upon this problem, either generally or specifically. There is no reason to suggest that they would object to this rule, however.

19. Hyde, *International Law, Chiefly as Interpreted and Applied by the United States*, I (Boston, 1947), p. 363.
20. Hsin Wu, "A Criticism of the Bourgeois International Law on the Question of State Territory," *KCWTYC*, 7:46 (1960).
21. See Oppenheim, *International Law*, I, 892–893; also arts. 48, 49 of the 1969 Convention on the Law of Treaties, *ILM*, 8:697–698, no. 4 (July 1969).
22. See Triska and Slusser, *The Theory, Law and Policy of Soviet Treaties* (Stanford, 1962), p. 54.

The Treaty-Making Process

The Communist Chinese process of making bilateral treaties usually follows a series of steps including the issuance of full powers, negotiation, signature, and ratification.[23] In general, the process does not differ from the procedures generally accepted by other states.

Once a state has decided to commence negotiations with another state over a particular treaty, the first step is to appoint a representative to conduct the negotiations. The representative of the state is usually provided with a formal instrument issued by the head of state, by the premier, or by the minister of foreign affairs, showing his authority to negotiate the treaty. This instrument is called the "full powers," and the document bearing the full powers of the representative is called "credentials."

It is the practice of Communist China to issue credentials bearing full powers to its representatives when it is to negotiate a formal treaty – an authority usually referred to in the preamble of the treaty. For example, the preamble of the Treaty of Peace and Friendship between (Communist) China and Nepal, [24] signed on April 28, 1960, contains a statement which reads: "the plenipotentiaries, having examined each other's credentials and found them to be in good and due form, have agreed . . . "

When the head of a state or government negotiates a treaty himself, full powers are not required.[25] Communist China apparently accepts this principle. When Liu Shao-Chi, Chairman of the People's Republic of China, signed several friendship treaties with other states, the preambles of those treaties made no reference to credentials, as they were not required.[26]

23. See the Decision on the Conclusion of Treaties and Other Agreements with Foreign Nations, adopted by the Government Administrative Council (renamed State Council after 1954) at its 102nd meeting, Sept. 14, 1951; "GAC Holds 102nd Meeting; Chang Han-fu Reports on Foreign Affairs," *SCMP*, 175:12 (Sept. 15–16, 1951). The Decision was revised Aug. 7, 1952. *SCMP* merely reports the passing of the decision; no text is available.

24. *TYC*, 10:13; English trans. in *Chung-hua jen-min kung-ho-kuo yu-hao t'iao-yüeh hui-pien* (Collection of friendship treaties of the People's Republic of China) (Peking, 1965), p. 12.

25. Jurists generally agree on this point, e.g., see art. 7, par. 2, 1969 Convention on the Law of Treaties; *ILM*, 8:683, no. 4 (July 1969).

26. See Communist China's friendship treaties with the Arab Republic of Yemen (June 9, 1964), *TYC*, 13:5; the Congo (B) (Oct. 2, 1964), *People's Daily*, Jan 11, 1965, p. 1; Mali (Nov. 3, 1964), *People's Daily*, Nov. 4, 1964, p. 1; Tanzania (Feb. 20, 1965), *People's Daily*, Feb. 21, 1965, p. 1.

When the premier of the State Council concludes a treaty with a foreign state, it is not clear whether the issuance of credentials bearing full powers is required. The practice of Communist China has not been consistent on this point.[27]

With respect to the requirement of credentials bearing full powers, the position of foreign minister is also unclear. Between 1949 and 1958, the premier of the State Council (called Government Administrative Council before 1954) in Communist China was concurrently the foreign minister, and sometimes it was difficult to tell whether he acted in his capacity as premier or foreign minister. Documents entitled "treaties" that were concluded after 1958 indicate that when a foreign minister signed a treaty, credentials bearing full powers were required. But since Communist China's foreign minister seldom signs documents entitled "agreement" with other states, his authority to sign an agreement without credentials bearing full powers remains uncertain.

Based on the information contained in the texts of many treaties, agreements, and other international documents in the official Compilations, all documents entitled "treaties," with the exception of several friendship treaties signed by the Chairman of the People's Republic of China or the Premier of the State Council, require the issuance of credentials bearing full powers prior to negotiation. The negotiation of most "agreements" does not seem to require the issuance of credentials bearing full powers.[28] Evidently this is also true of other documents, such as protocols, exchanges of notes, and semiofficial agreements.

In Communist China's practice, usually the negotiation of a bilateral treaty is conducted by representatives of or delegations from the contracting parties. It is an international practice that when credentials bear-

27. Several friendship treaties signed by Chou En-lai, premier of the State Council, do not refer to credentials in the preamble; *e.g.*, see preamble of Sino-Burmese Treaty of Friendship and Mutual Non-Aggression (Jan. 28, 1960), *TYC*, 9:44. But several friendship treaties signed by Chou do refer to credentials in the preamble; *e.g.*, preamble of Sino-Nepalese Treaty of Peace and Friendship (Apr. 28, 1960), *TYC*, 10:13. Usually a treaty preamble refers to "credentials" instead of "full powers," but the French text of the Sino-Cambodian Treaty of Amity and Mutual Non-Aggression (Dec. 19, 1960) refers to "pleins pouvois" (full powers), while the Chinese text refers to "ch'üan-ch'üan cheng-shu" (credentials); in the Collection, n. 24 above pp. 33, 38.

28. The preambles of some agreements do refer to credentials; *e.g.*, Cultural Cooperation Agreement between [Communist] China and Poland (Apr. 3, 1951), *TYC*, 1:102, or Agreement between India and [Communist] China on Trade and Intercourse between the Tibetan Region of China and India (Apr. 29, 1954), *TYC*, 3:1; *UNTS*, 299:57.

ing full powers are required in negotiating a treaty, the representatives should first examine each other's credentials. Communist China follows this practice.

At the conclusion of the negotiations, Communist China and the other negotiating state usually issue a communiqué mentioning the period of negotiation, the names of the negotiating officers on both sides, and the title of the treaty signed. Occasionally a communiqué is issued when the preliminary negotiations are concluded. For instance, in the course of negotiating the Sino-Indonesian Dual Nationality Treaty, two communiqués were issued: the first on December 29, 1954, announcing the conclusion of the preliminary negotiation;[29] the other on April 22, 1955, when the treaty was signed.[30]

Communist China believes that no information should be unilaterally disclosed to the public while negotiations are in progress. In 1966, when Premier Castro unilaterally disclosed the contents of Sino-Cuban trade negotiations, a responsible official of the Communist Chinese Ministry of Foreign Trade commented:

> According to the usual practice of trade negotiations between China and Cuba, preliminary discussions on trade between the two countries for 1966 are first held in Peking, and then the Chinese Government will send a delegation to Havana for the formal signing of the annual protocol. At the moment, the delegation of the Cuban Ministry of Foreign Trade is still in Peking, the preliminary trade negotiations between the two sides are going on, and the annual protocol has not yet been finally signed. If the Cuban Government had different ideas or demands, it can very well raise them for discussion with the Chinese Government. But instead of doing so, Prime Minister Castro has taken a step which is extraordinary in normal state relations. At the Havana mass meeting celebrating the 7th anniversary of the liberation of Cuba [held on January 2, 1966], he unilaterally and untruthfully made public contents of the preliminary trade negotiations now going on between the governmental departments concerned of the two countries. We cannot but feel regret at this.[31]

The common practice of most countries is that, when negotiations have been concluded and the treaty or agreement embodied in the prop-

29. *WCC*, 3:233; English trans. in "Sino-Indonesian Joint Communiqué on Preliminary Negotiations on Dual Nationality," *SCMP*, 957:9 (Dec. 30, 1954).
30. *WCC*, 3:275; English trans. in "Communiqué on Sino-Indonesian Talk on Dual Nationality," *SCMP*, 1033:27 (Apr. 23–25, 1955).
31. "Facts on Sino-Cuban Trade," *PR*, 9:22–23, no. 3 (Jan. 14, 1966).

er form, the representatives sign the document. The effect of the signature depends upon whether the signed treaty or agreement requires ratification; those that do not come into force upon the date of signature or upon some other date if provided in the document. In this respect, Communist China's practice on signature does not seem to differ from that of other states.

Occasionally, a representative of a contracting party signs a treaty or agreement ad referendum or by initialing; Western jurists generally agree that this does not bind a government.[32] Communist Chinese theory seems similar. For example: "It should be pointed out that initialing and formal signature are two completely different matters. Ordinarily, the representatives only put down an abbreviation of their names when initialing a draft treaty; such an act merely indicates that the initialing representative temporarily signed and confirmed the formal consent of his government to the draft, and he must await the final instruction of his government before he can formally sign the draft."[33]

Generally speaking, ratification is the final confirmation given by the parties to an international treaty concluded by their representatives, though today many treaties or agreements do not require ratification. The appropriate procedures are left to the internal law of the state concerned.

In Communist China, before the promulgation of the 1954 Constitution, treaties or agreements requiring ratification were submitted to the Central People's Government Committee, which was elected by the Chinese People's Political Consultative Conference.[34] Since 1954, a treaty or agreement which requires ratification must first be passed by the State Council, which submits it to the Standing Committee of the National

32. *E.g.*, see Judge Fitzmaurice's report on the law of treaties in *Yearbook of the International Law Commission 1956,* 2:101, U.N. Doc. A/CN. 4/ ser. A/1956/ Add. 1.

33. Wei Liang, "Looking at the So-called McMahon Line from the Angle of International Law," *KCWTYC*, 6:48 (1959).

34. See arts. 6, 7 of Organic Law of the Central People's Government of the People's Republic of China, adopted by First Plenary Meeting of the Chinese People's Political Consultative Conference, Sept. 27, 1949; *FLHP*, 1:2. It appears that after 1952 treaties or agreements requiring ratification were passed by Government Administrative Council and then submitted to Central People's Government Committee for ratification; *e.g.*, Sino-[North] Korean Agreement on Economic and Cultural Cooperation (Nov. 23, 1953), *TYC*, 2:6, was first passed by Government Administrative Council at 195th meeting, Nov. 26, 1953, then ratified by Central People's Government Committee at 29th meeting, Dec. 9, 1953; see *People's Daily*, Nov. 28, 1953, p. 1; Dec. 10, 1953, p. 1.

People's Congress for a decision as to whether ratification should be granted. Finally, the Chairman of the People's Republic of China formally ratifies the treaty or agreement in accordance with the decision of the Standing Committee of the National People's Congress.[35]

At the same time Communist China established its current formal process of ratifying treaties or agreements it adopted a simplified process for "approval" of some agreements. An agreement requiring "approval" is submitted to the State Council only; it does not have to be submitted to the Standing Committee of the National People's Congress.

As a result of the Cultural Revolution, the Standing Committee of the National People's Congress has ceased to function since mid-1966. The Chairman of the People's Republic of China, Liu Shao-chi, has also ceased to exercise his functions since late 1966. Subsequently, Communist China has not concluded any treaty or agreement which, in light of past practice, requires "ratification"; so it is not clear which organs should now ratify treaties or agreements.

In the early stages of the Cultural Revolution, the State Council's "approval" function was apparently not affected, and the Council approved several agreements in early 1967.[36] Since then, however, no agreement has been approved by the State Council. Since Communist China's practice concerning approval is very flexible, it is not clear whether this fact indicates that the approval function of the Council has been affected by the Cultural Revolution.[37]

There is no general rule of international law prescribing what kinds of treaties or agreements require ratification. This question is determined by

35. Art. 31 of 1954 Constitution of Communist China provides, inter alia, that "the Standing Committee of the National People's Congress exercises the following functions and powers . . . (2) to decide on the ratification . . . of treaties concluded with foreign states," art. 41 that "the Chairman of the People's Republic of China . . . ratifies treaties concluded with foreign states"; *FKHP*, 1:14, 17. On Sept. 6, 1970, the 2nd Session of the 9th Central Committee of the Chinese Communist Party prepared a revised draft of the "Constitution of the People's Republic of China"; art. 18 contains a provision similar to the provision of art. 31 cited above; *Issues & Studies* (Taipei), 7:92, no. 3 (December 1970).

36. *E.g.*, on Mar. 14, 1967, the State Council approved the Sino-Mauritanian Agreement on Economic and Technical Cooperation of Feb. 16, 1967; "China's State Council Approves Sino-Mauritanian Agreements," *SCMP*, 3907:38 (Mar. 29, 1967).

37. According to the official Compilation, 15 economic aid, loan, or technical cooperation agreements were concluded during the period 1949–1964; only one was ratified, 2 were approved. Between early 1967 and early 1969 Communist China concluded several such agreements; none was approved by the State Council; *e.g.*, see Sino-Pakistanian Agreement on Economic and Technical Cooperation (Dec. 26, 1968), "China and Pakistan Sign Economic and Technical Co-operation Agreement," *PR*, 12:26, no. 2 (Jan. 10, 1969).

the internal law of the contracting states and by the provisions of the treaty or agreement itself. On October 16, 1954, the Standing Committee of the National People's Congress passed the following resolution concerning this question:

> 1. The following types of treaties concluded by the People's Republic of China with foreign states are to be ratified in accordance with the provisions of Article 31, Section 12, and Article 41 of the Constitution of the People's Republic of China: peace treaties; treaties of nonaggression; treaties of friendship, alliance and mutual assistance; and all other treaties, including agreements, which contain stipulations that they are to be submitted for ratification.
> 2. Agreements and protocols which do not fall under the above categories are to be approved by the State Council.[38]

The literal sense of the second paragraph implies that all agreements or protocols which do not require ratification should be "approved" by the State Council. But the practice of Communist China does not support such an interpretation; many agreements or protocols do not require either "ratification" or "approval."

With three exceptions, all documents entitled "treaties" that were concluded between 1949 and 1964 were ratified. Two of the three treaties not subjected to ratification were signed by the Chairman of the People's Republic of China and the heads of states of the other contracting parties. However, before the Chairman signed these two treaties, the drafts were first passed by the State Council and then sent to the Standing Committee of the National People's Congress for "examination." The Standing Committee examined the drafts and afterwards authorized the Chairman to sign the treaties.[39] The other treaty not subjected to ratification was signed by the Foreign Minister of Communist China. As in the two cases just mentioned, the draft of this treaty was first passed by

38. *FKHP*, 1:207. Communist Chinese process of ratifying or approving treaties or agreements is similar to Soviet process: under the Soviet law of Aug. 20, 1938, on the procedure for ratification and denunciation of international treaties, treaties are to be ratified by the Presidium of the Supreme Soviet. The law provides that peace treaties, treaties re mutual defense against aggression, and treaties of mutual nonaggression are subject to ratification. Treaties at the conclusion of which the parties have agreed to a subsequent ratification are also subject to ratification; see "On the Procedure for the Ratification and Denunciation of International Treaties of the USSR," *Soviet Statutes and Decisions*, 3:56–57, no. 4 (Summer 1967). In addition to ratification, some treaties or agreements are subject to "confirmation" by the Council of Ministers only; see Triska and Slusser, *Soviet Treaties*, p. 79.

39. The two treaties are: (1) Sino-Nepalese Boundary Treaty (Oct. 5, 1961), *TYC*, 10:45, and (2) Sino-Yemen Arab Republic Friendship Treaty (June 9, 1964), *TYC*, 13:5. The draft of the first treaty was passed by the State Council at its 113th plenary session, Oct. 5, 1961. It was immediately submitted for "examination" to the Standing Committee of the National People's Congress, which authori-

the State Council and then "examined" by the Standing Committee of the National People's Congress which, after examination, appointed the Foreign Minister to sign the treaty.[40]

Communist Chinese practice with respect to whether other types of international documents such as agreements and protocols require ratification, approval, or just signature does not seem consistent. Annual executive plans or protocols for implementing cultural or scientific cooperation agreements do not usually require approval or ratification, though there is at least one exception.[41]

Table 2, on signature, approval, and ratification, is based upon a study of bilateral treaties, agreements, protocols, and other documents contained in the 1949–1964 official Compilation. It should be noted that some agreements, while explicitly referring to "ratification," were in fact subject to "approval" only.[42]

---

zed Chairman Liu Shao-chi to sign the treaty at its 43rd meeting, held that day; "State Council Adopts Draft Sino-Nepalese Boundary Treaty at 113th Meeting," *SCMP*, 2596:1 (Oct. 11, 1961). The draft of the second treaty was passed by the State Council at its 146th plenary session, June 9, 1964. It, too, was immediately submitted for examination to the Standing Committee, and Chairman Liu Shao-chi was authorized to sign it at the 119th meeting held the same day; "State Council Holds 146th Meeting–Approves (Draft) Treaty of Friendship Between China and Yemen," *SCMP*, 3237:9 (June 12, 1964), and "NPC Standing Committee Holds 119th Meeting--Decides that Chairman Liu Shao-ch'i Should Sign China-Yemen Friendship Treaty," *ibid.*

40. Sino-Afghanistan Boundary Treaty (Nov. 11, 1963), *TYC*, 12:122. Draft of the treaty was passed by the State Council at its 136th plenary session, Oct. 23, 1963, and submitted for examination to the Standing Committee of the National People's Congress, which examined the treaty and appointed Foreign Minister Ch'en Yi to sign it at the 106th meeting, Nov. 9, 1963; "State Council Passes at 136th Plenary Session Border Treaty with Afghanistan," *SCMP*, 3088:1 (Oct. 28, 1963); "NCP Standing Committee Holds 106th Meeting," *SCMP*, 3100:1 (Nov. 14, 1963).

41. On Nov. 21, 1957, Communist China and the Soviet Union "concluded" discussions on the 1958 Executive Plan for implementing the Sino-Soviet Cultural Cooperation Agreement of July 5, 1956, *TYC*, 5:152; "China, U.S.S.R. Discuss 1958 Cultural Cooperation Plan," *SCMP*, 1659:39–40 (Nov. 26, 1957). On Jan. 9, 1958, the Plan was "ratified" by the State Council at its 68th session; *People's Daily*, Jan. 10, 1958, p. 1; "State Council Holds 68th Meeting," *SCMP*, 1691:1 (Jan. 15, 1958).

42. *E.g.*, art. 8 of Sino-Syrian Payment Agreement, signed Nov. 30, 1955, explicitly provided that the agreement was to be "ratified," but in fact it was only "approved" by the Communist Chinese State Council; *TYC*, 4:122; see also art. 8, Sino-Cuban Agreement on Cultural Cooperation, July 23, 1960, *TYC*, 10:389. A few agreements provide explicitly or implicitly that they are subject to either "approval" or "ratification." *E.g.*, art. 4 of Sino-Syrian Accord on Amending Trade Agreement and Payment Agreement between the Two Countries (July 3, 1957) provided that it would be subject to "ratification" or "approval" in accordance with legal processes of the contracting parties; it was in fact approved by the Communist Chinese State Council, *TYC*, 6:162, 163n. Art. 8 of Sino-Indonesian Agreement on Economic and Technical Cooperation (Oct. 11, 1961) provided that it was to be executed in accordance with the legal processes of each government; it was in fact approved by the Communist Chinese State Council, *TYC*, 10:247.

Table 2. Signature, approval, and ratification, as classified in the official Compilation

Type of agreement	Title of agreement	Number of agreements requiring signature only	Number of agreements requiring approval	Number of agreements requiring ratification	Observation
Political					
(1) Friendship	Treaty	2	–	14	–
	Exchange of notes	1	–	–	Abolishing Sino-Afghanistan Friendship Treaty of 1944
(2) Joint statements, communiqués, and declarations	Joint statements, communiqués, and declarations	102	–	–	–
(3) Other	Exchange of notes	4	–	–	–
	Agreement	–	–	5	–
	Joint communiqué	1	–	–	1954 Soviet withdrawal from Port Arthur
	Announcement	1	–	–	1955 Sino-American announcement on the return of civilians
Legal					
(1) Consular relations	Treaty	–	–	3	–
(2) Nationality	Treaty	–	–	1	–
	Exchange of notes	1	–	–	–

	Measures for implementation	1	–	–	–
Boundary	Treaty	2	–	2	–
	Agreement	1	–	2	–
	Exchange of notes	2	–	–	–
	Protocol	1	1	–	–
	Press communiqué	4	–	–	–
	Joint communiqué	1	–	–	–
Boundary problems (use of boundary rivers, etc.)	Agreement	1	–	–	–
	Protocol	3	–	–	–
Economic					
(1) Commerce and navigation	Treaty	–	–	7	–
	Exchange of notes	4	–	–	–
(2) Economic aid; loan and technical cooperation	Agreement	12	1	2	–
	Protocol	1	–	–	–
	Exchange of notes	4	–	–	–
(3) Trade and payment	Agreement	97	18	6	Not including 4 agreements which were to be continued by exchange of notes
	Protocol	77	–	–	
	Exchange of notes	36	–	–	
	Accord	–	3	–	–
	Communiqué	1	–	–	–
(3a) Semiofficial	Accord	3	–	–	–
	Protocol	4	–	–	–
	Exchange of notes; letters	4	–	–	Including 1 "official letter"
	Joint announcement	3	–	–	–
	Agreement	4	–	–	–
(4) General con-	General conditions for	71	–	–	–

Table 2, continued

Type of agreement	Title of agreement	Number of agreements requiring signature only	Number of agreements requiring approval	Number of agreements requiring ratification	Observation
ditions for the	the delivery of goods				
delivery of goods	Protocol	11	–	–	–
	Exchange of notes	4	–	–	–
(4a) General conditions for the delivery of goods (semiofficial)	General conditions for the delivery of goods	2	–	–	–
(5) Registration of trademark	Exchange of notes	4	–	–	–
(6) Other	Exchange of notes	7	–	–	–
	Agreement	6	–	–	–
	Protocol	1	–	–	–
	Communiqué	1	–	–	–
Cultural					
(1) Cultural	Agreement	1	12	16	–
cooperation	Protocol	1	–	–	–
(2) Broadcasting	Agreement	27	–	–	–
and television	Protocol	2	–	–	–
cooperation	Exchange of notes	1	–	–	–
(3) Exchange of	Agreement	2	–	–	–
students	Protocol	1	–	–	–
(4) Other	Communiqué	1	–	–	–

Science and Technology	Agreement	15	1	1	Approved or ratified by Standing Committee of Chinese Academy of Sciences
	Protocol	3	–	1	
	Accord	2	4	1	
	Exchange of notes	5	–	–	
	Joint communiqué	2	–	–	–
Agriculture	Agreement	6	1	1	–
	Exchange of notes	1	–	–	–
Fishery	Communiqué	1	–	–	–
	Agreement	2	–	–	–
Semiofficial	Exchange of memoranda	2	–	–	–
	Exchange of letters	2	–	–	–
	Exchange of notes	5	–	–	–
	Protocol	1	–	–	–
Health and Sanitation	Agreement	2	1	1	–
	Protocol	1	–	–	–
Post and telecommunication	Agreement	26	2	8	–
	Protocol	5	–	–	–
	Exchange of notes	1	–	–	–
Communication and transportation					
(1) Railway	Agreement	2	–	–	–
	Protocol	1	–	–	–
(2) Air transportation	Agreement	9	–	–	–
	Accord	1	–	–	–
	Protocol	1	–	–	–
(3) Water transportation	Agreement	3	2	1	–
	Exchange of notes	3	–	–	–
(4) Highway	Agreement	1	–	–	–

International law does not determine when a treaty or agreement enters into force; this is a problem for the contracting parties to decide. A treaty can specifically provide that it will come into force upon the date of signature, on a specific date, after the exchange of ratification, or at some other time.

In Communist Chinese practice, formal "treaties" requiring ratification are usually effective on the exchange of ratifications.[43] This is also usually the practice of Western states. The practice with respect to "agreements" varies. Agreements requiring ratification have come into force upon a specific date, upon the date of signature, upon ratification, upon notification of ratification, and upon exchange of ratification.[44] A few agreements have been effective on the date when the contracting parties exchanged notes concerning the entry into force, or on the date of signature but with confirmation by exchange of notes required.[45]

Agreements or protocols requiring approval have usually come into force upon the date of approval, the notification of approval, or the exchange of the notes of approval.[46] Many agreements or protocols specifi-

43. The only exception seems to be the Sino-Soviet Treaty of Friendship and Alliance (Feb. 14, 1950), which provided in art. 6 that it would come into force upon ratification; *TYC*, 1:2.

44. Examples: the Sino-[East] German Agreement on Exchange of Goods and Payment in 1953 (Apr. 30, 1953); it came into force Jan. 1, 1953, and was subject to ratification by both parties (art. 8); *TYC*, 2:161.

The Sino-Soviet Agreement on Soviet Loan to China (Feb. 14, 1950); art. 5 provided that it was to come into force upon signature, though ratifications were exchanged in Peking later; *TYC*, 1:46.

Art. 4 of Sino-Polish Agreement on Technical Science Cooperation (July 20, 1954) provided that it was to come into force upon the date of ratification; *TYC*, 3:172.

Art. 7 of Sino-Czechoslovakian Agreement on Cultural Cooperation (Mar. 27, 1957) provided that it would come into force upon the notification of ratifications. It was ratified by the Chairman of the People's Republic of China on May 12, 1957, and notification was sent to the Czechoslovakian government May 16, 1957. It was ratified by the President of Czechoslovakia on May 31, 1957, and notification of ratification was sent to the Communist Chinese government Aug. 9, 1957. The agreement came into force Aug. 9, 1957; *TYC*, 6:214.

Art. 4 of Sino-Burmese Agreement on Boundary Question (Jan. 28, 1960) provided for effectiveness on exchange of ratifications; *TYC*, 9:68.

45. Examples: art. 8 of Sino-Indonesian Trade Agreement (Nov. 30, 1953) provided for effectiveness on exchange of notes between the two countries; *TYC*, 2:34.

Art. 12 of Sino-Indonesian Payment Agreement (Sept. 1, 1954) provided for effectiveness on signature, but with confirmation by exchange of notes between the two countries; *TYC*, 3:42.

46. Examples: art. 9 of Sino-Egyptian Trade Agreement (Aug. 22, 1955) provided for effectiveness on approval by both countries; *TYC*, 4:124.

Art. 12 of Sino-Iraqi Agreement on Cultural Cooperation provided for effectiveness on reciprocal notification of approval; *TYC*, 8:135.

cally provide that they will be enforced upon the date of signature.[47] But many are silent on the questions of ratification, approval, or effective date; presumably they come into force upon signature.

In agreements or protocols that come into force upon a specific date, the peculiar practice of Communist China is that many of them have actually come into force retroactively up to several months before the date of signature.[48]

Other documents, such as those entitled "exchange of notes," "declaration," "communiqué," "joint announcement," and "accord," usually are silent on the question of effective date; presumably they come into force upon the date of signature.

Information concerning the signature, approval, and ratification of treaties and other agreements is generally reported by the official New China News Agency and printed in the *People's Daily*, the official newspaper of the Chinese Communist Party. Occasionally the New China News Agency and the *People's Daily* give the complete text of a treaty, but more frequently they simply mention or summarize its contents. Since 1955 the official Collection of Laws and Regulations of the People's Republic of China (*FKHP*) has included the texts of certain treaties relating to questions of nationality, boundaries, commerce and navigation, and so on.

In 1957 Communist China's Ministry of Foreign Affairs published its first volume of the Compilation (*TYC*). By 1965 thirteen volumes had been published, containing most of the treaties, agreements, protocols, exchanges of notes, and other commitments concluded by Communist China between 1949 and 1964. Except for the first two volumes, which

---

Art. 5 of Sino-[East] German Agreement on Cultural Cooperation provided that it would become effective upon the date of exchange of notes of approval; *TYC*, 4:210.

47. *E.g.*, almost all agreements on cooperation in broadcasting and television have come into force upon signature. Effectiveness on signature may be provisional, in which case later confirmation is necessary. Art. 9 of Sino-Swedish Trade Agreement (Nov. 8, 1957) provided that it would become provisionally effective on the date of signature and formally effective upon exchange of notes between the two countries; *TYC*, 6:183.

48. Many trade or payment agreements or protocols concluded in the course of a year were made effective retroactively from the beginning of that year; *e.g.*, 10 of 18 concluded Jan. 9–Nov. 11, 1961, were made effective retroactively from Jan. 1, 1961.

The practice of making agreements retroactively effective is not limited to trade or payment agreements. The Protocol between Communist China and Ghana for the Dispatch of Chinese Military Experts to Ghana was signed Aug. 5, 1965, but made retroactively effective from Sept. 30, 1964; Ghana Ministry of Information, *Nkrumah's Subversion in Africa* (Accra, 1966), p. 56.

contain treaties concluded in the periods 1949–1951 and 1952–1953 respectively, each volume is comprised of treaties concluded in a given year. Some important semiofficial agreements are included in the appendix.

The official Compilation does not include some agreements of minor importance, such as annual executive plans or protocols for implementing cultural or scientific cooperation agreements. Occasionally, agreements of considerable importance are excluded. For instance, the February 14, 1950, exchange of notes between Communist China and the Soviet Union on abolition of the 1945 Sino-Soviet Treaty of Friendship and on the recognition of Outer Mongolia was not included.[49]

The United Nations Charter provides in Article 102 that "every treaty and every international agreement entered into by any Member of the United Nations after the . . . Charter comes into force shall as soon as possible be registered with the Secretariat and published by it." Hitherto, only a small number of treaties concluded by Communist China have been registered with the United Nations; they were registered not by Communist China but by its treaty partners. However, this fact should not be interpreted to mean that Communist China rejects the system of treaty registration. A quite plausible explanation of its reluctance to register treaties with the United Nations is that it has not been allowed to participate in the activities of the United Nations as the "legitimate representative" of China.

A state may participate in a conference to conclude a multilateral treaty, or it may adhere to or accept an existing multilateral treaty. In signing, ratifying, adhering to, or accepting a multilateral treaty, a state may make reservations to one or more of its articles. Communist China has become a party to a small number of multilateral treaties; its practice with respect to their conclusion does not seem to differ from that of other states.

49. The exchange of notes was registered with the U.N. Secretariat by the Soviet Union; *UNTS*, 226:16. Occasionally, Communist China's diplomatic statements or notes reveal the existence of some agreements not reported by *NCNA*, *People's Daily*, or the official Compilation at the time of concluding those agreements; *e.g.*, on Aug. 15, 1963, the spokesman of the Communist Chinese Ministry of Posts and Telecommunication made a statement urging India to honor her postal agreements with Communist China and accusing India of violating a postal agreement concluded with Communist China in 1960; *People's Daily*, Aug. 16, 1963, p. 3; English

Communist China has taken part in several international conferences which eventually adopted multilateral treaties. The 1956 Railway Cooperation Conference among socialist countries and the 1962 Geneva Conference on the Question of Laos are two examples. One special problem which often precludes Communist Chinese attendance at international conferences is its strong opposition to concurrent participation by the Republic of China.[50] Communist China has never participated in an international conference at which the Republic of China has been represented. It should be noted that Communist China has been excluded from many treaty-making conferences sponsored by the United Nations, as participation in these conferences has been limited to member states of the United Nations and specialized agencies or states especially invited by the United Nations General Assembly.

The nature of a conference determines whether participation requires the issuance of credentials bearing full powers to representatives.[51] As for Chinese ratification or approval of multilateral treaties, the actual cases are few, and one can hardly generalize from them. Only the four Geneva Red Cross Conventions have been formally ratified by the Chairman of the People's Republic of China in accordance with the decisions of the Standing Committee of the National People's Congress;[52] two agreements were approved by the State Council;[53] all other agreements

---

trans. "India Urged to Honor Postal Agreement with China," in *SCMP*, 3043:19 (Aug. 20, 1963). This agreement had not been reported previously. On Feb. 10, 1967, a representative of Communist China's Foreign Ministry protested to the Soviet Embassy in China against the latter's violation of several agreements on the exemption of visas concluded since 1956; *People's Daily*, Feb. 11, 1967, p. 5. These agreements were not reported previously.

50. For a collection of relevant Communist Chinese documents and articles on "two Chinas," see *Oppose the New U.S. Plots to Create "Two Chinas"* (Peking, 1962).

51. *E.g.*, the Agreement Concerning Cooperation in the Field of Research in Fishery, Oceanology, and Limnology in the Western Part of the Pacific Ocean signed (June 12, 1956) by the Soviet Union, Communist China, North Korea, and North Vietnam did refer to credentials in its preamble; *TYC*, 5:169. See also the Agreement on Establishing a Joint Institute for Nuclear Research (Mar. 26, 1956), *TYC*, 10:408. But the Agreement Concerning Collaboration in Shipwreck, Salvage Services, Saving Human Lives and Aiding Ships and Aircraft on Seas signed (July 3, 1956) by the Soviet Union, Communist China, and North Korea did not refer to credentials; *TYC*, 5:199.

52. *TYC*, 5:231 (land warfare), 255 (sea warfare), 332 (prisoners of war), 403 (civilians).

53. See the Regulations on Railway Cooperation Organization (June 28, 1956), *TYC*, 6:314; and the Accord on Post and Telecommunication Cooperation Organization Among Socialist Countries (Dec. 16, 1957), *TYC*, 6:271.

either explicitly provided for entry into force upon signature[54] or were silent on the question.[55]

Communist China has thus far adhered to only one multilateral treaty, the 1929 Warsaw Convention on International Carriage by Air.[56] The decision on adherence was adopted by the Standing Committee of the National People's Congress on June 5, 1958; the notification of adherence was sent to Poland, the depository state, on July 19, 1958.[57] Communist China has so far accepted only the 1948 International Regulations for Preventing Collision at Sea.[58] Here again, decision on acceptance was adopted by the Standing Committee of the National People's Congress on December 23, 1957.[59]

No Communist Chinese writers have dealt with the question of reservation, but Communist China has at times made reservations to multilateral conventions. For example: in 1956, when ratifying the Geneva Convention Relative to the Treatment of Prisoners of War,[60] signed in behalf of China by Nationalist China on August 12, 1949, it made the following reservations:

> Regarding Article 10 . . . the People's Republic of China will not recognize as valid a request by the Detaining Power of prisoners of war to a neutral State or to a humanitarian organization, to undertake the functions which should be performed by a Protecting Power, unless the consent has been obtained of the government of the State of which the prisoners of war are nationals. Regarding Article 12, the People's Republic of China holds that the original Detaining Power which has transferred prisoners of war to another Contracting Power, is not for that reason freed from its responsibility under the Convention while such prisoners of war are in the custody of the Power ac-

54. *E.g.*, art. 5 of Agreement Concerning Direct Rail Communication between the Soviet Union, Communist China, and Mongolia (Sept. 15, 1952) explicitly provided for effectiveness on signature; *TYC*, 2:241.

55. *E.g.*, Final Declaration of the Geneva Conference (July 21, 1954), *TYC*, 3:14.

56. Hudson, *International Legislation* 5:100 (Washington, 1936).

57. *FKHP*, 7:338; *TYC*, 7:195n. The editor of *TYC* reported that Communist China on July 31, 1953, "adhered" to both the Agreement Concerning International Freight Communication and the Agreement Concerning International Passenger Communication. But both agreements were signed by Communist Chinese representatives, with representatives of other Communist Countries, on that day in Moscow, and the agreements came into force Jan. 1, 1954; *TYC*, 2:284 (passenger), 341 (freight).

58. *TYC*, 6:294; *UNTS*, 191:20.

59. *TYC*, 6:313.

60. *TYC*, 5:255; *UNTS*, 75:135.

> cepting them. Regarding Article 85, the People's Republic of China is not bound by Article 85 in respect of the treatment of prisoners of war convicted under the laws of the Detaining Power in accordance with the principles laid down in the trials of war crimes or crimes against humanity by the Nuremberg and the Tokyo International Military Tribunals.[61]

The decision on reservation was made by the Standing Committee of the National People's Congress at the time it decided to grant ratification to the Convention.[62]

Regarding the specific rule that Communist China believes should govern reservations, no clear answer can be given: neither the official statements nor the writings of Communist Chinese discuss the matter. In view of Communist China's strong emphasis on state sovereignty as the most fundamental principle of international law,[63] it would probably accept the present Soviet position that a state has the sovereign right to make reservations to a multilateral treaty without the consent of other contracting parties. But in that case the objecting states would be entitled to regard themselves as not bound by the reserved provisions of the convention vis-à-vis the reserving state, or to consider the reserving state as not a party to the convention.[64]

When signing, ratifying, acceding to, or accepting a treaty, general international practice recognizes that a state may make explanatory declarations, or statements of intention or recognition, that do not possess the character of reservations.[65] Communist China also follows this practice. Thus, when ratifying the International Regulations for Preventing Col-

61. *UNTS*, 260:442; *TYC*, 5:332.

62. *FKHP*, 4:229; *TYC*, 5:332.

63. *E.g.*, see Ying T'ao, "A Criticism of Bourgeois International Law concerning the Question of State Sovereignty," *KCWTYC*, 3:47 (1960); Yang Hsin and Chen Chien, "Expose and Censure the Imperialist's Fallacy concerning the Question of State Sovereignty," *CFYC*, 4:6 (1964).

64. See Triska and Slusser, *Soviet Treaties*, pp. 87–88. Communist China apparently does not consider that consent of other contracting parties is required in making a reservation; thus, despite the reservations of many parties to the Geneva Conventions when ratifying or adhering to them, Communist China did not bother to express approval or disapproval of the reservations. In a Chinese text of the conventions published by Communist China's Foreign Ministry, the list of contracting parties includes all ratifying or adhering states without differentiating those which made reservations; Chung-hua jen-min kung-ho kuo wai-chiao pu pien, *I-chiu-szu-chiu nien pa-yüeh shih-erh jih jih-nei-wa kung-yüeh* (The Geneva Conventions of 12 August 1949) (Peking, 1958), pp. 178–182.

65. For a study of this question, see Chiu, "Reservations and Declarations Short of Reservations to Treaties," *Journal of Social Sciences*, 15:71 (Taipei, July 1965).

lision at Sea on June 10, 1948, Communist China made the following statement in connection with its deposit of ratification on January 29, 1959:

> The Government of the People's Republic of China makes the following reservation: that these regulations do not have binding force on non-engine-driven ships belonging to the People's Republic of China.
> The Government of the People's Republic of China is the sole legitimate Government representing the Chinese People. The Regulations for Preventing Collisions at Sea shall naturally apply in the whole territory of China including Taiwan and all other islands belonging to China.[66]

The second part of the above statement does not constitute a reservation, because it does not "purport to exclude or to vary the legal effect of certain provisions of the treaty in their application to that state."[67]

Treaties and Third States

As a general rule, a treaty may not impose obligations or confer rights on a third state. This principle, expressed in the Latin maxim *pacta tertiis nec nocent nec prosunt*, has been invoked by Communist Chinese writer Yü Fan in discussing the question of Chinese "suzerainty" over Tibet.[68]

Yü points out that in the history of British "aggression" against Tibet, Britain tried several times between 1904 and 1914 to create a fiction of "Chinese suzerainty over Tibet" in order to negate the real sovereignty of China in the area. The British argument was composed largely of references to treaties, one of them the Convention between Britain and Russia relating to Persia, Afghanistan, and Tibet, signed at St. Petersburg on August 31, 1907.[69] Article 2 of the Convention read: "In conformity with the admitted principle of the *suzerainty of China over Tibet*, Great Britain and Russia engage not to enter into negotiations with Tibet except through the intermediary of the Chinese Government." (Emphasis added.)

66. *TS*, no. 39 (1959), Cmd. 727 (United Kingdom).
67. Art. 2 of Draft Articles on the Law of Treaties adopted by the U.N. International Law Commission in January 1966; U.N. General Assembly, *Official Records*, 21st sess., supp. no. 9 (A/6309/rev. 1, 1966).
68. "Speaking about the Relationship between China and the Tibetan Region from the Viewpoint of Sovereignty and Suzerainty," *People's Daily*, June 5, 1959, p. 7.
69. *BFSP*, 100:555 (1906–1907).

Commenting on the binding force of this Convention upon China, Yü wrote: "This Convention, however, was concluded between Britain and Russia and can only be effective between Britain and Russia. According to the principle of *'pacta tertiis nec nocent nec prosunt'* of the law of treaties, since China was not a contracting party, this Convention naturally cannot bind China. How can it impose a fictitious 'suzerainty' on the relationship between China and Tibet by a convention of which China was ignorant?"

Although as a general rule a treaty may not confer rights upon a third state, many Western scholars of international law express the view that "if a treaty stipulates a right for third states, and they make use of such a right, they thereby acquire a legal right for themselves."[70]

In its diplomatic exchange with India concerning the Sino-Indian border, Communist China invoked this principle in its note of December 26, 1959, to India. The note said in part:

> It must . . . also be pointed out that it is beyond question that Britain had no right to conduct separate negotiations with Tibet. Indeed, the Chinese Government made repeated statements to this effect; as to the British Government, it too was strictly bound by the 1907 agreement on Tibet concluded between it and the old Russian Government not to enter into negotiations with Tibet except through the intermediary of the Chinese Government. Therefore, judging by this treaty obligation alone which was undertaken by the British Government, the secret exchange of letters in 1914 between the British representative and the representative of the Tibet local authorities behind the back of the Chinese Government is void of any legal validity.[71]

Ironically, the 1907 treaty invoked here against Britain by Communist China is the same treaty which Yü Fan, relying on the maxim *pacta tertiis nec nocent nec prosunt*, considered not binding upon China.

Some multilateral treaties declaratory of established customary international law obviously apply to third parties; but legally speaking, the third parties are bound not by the treaty but by the customary rules.[72] A Communist Chinese article denouncing the alleged use of "poisonous gas" by the United States in Vietnam discussed this issue.[73] The writer

70. Oppenheim, *International Law*, I, 8th ed. (London, 1955), 927.
71. *WCC*, 6:129–130; English trans. in *The Sino-Indian Boundary Question*, rev. ed. (Peking, 1962), p. 60.
72. *E.g.*, see Starke, *Introduction to International Law*, (London, 1967), p. 345.
73. Fu Chu, "American Imperialism's Use of Poisonous Gas in South Vietnam is a War Crime in Flagrant Violation of International Law," *People's Daily*, Apr. 3, 1965, p. 3.

argued that although the United States had not ratified the 1925 "Geneva Protocol Relating to the Prohibition of the Use in War of Asphyxiating, Poisonous or Other Gases, and of Bacteriological Methods of Warfare,"[74] the principle provided in that Protocol was "universally recognized." Accordingly, the author concluded, the United States "cannot escape criminal responsibility for its violation" of the principle of international law prohibiting the use of poisonous gas.

Many commercial treaties or trade agreements concluded among states contain a stipulation usually referred to as the "most-favored-nation" clause. The wording of this clause is by no means the same in all treaties or agreements. Generally speaking, however, it provides that all favors which either of the contracting parties has granted in the past, or will grant in the future, to any third state must be granted to the other party.[75] Thus, a commercial treaty or trade agreement conceding more favorable conditions has an effect upon all such states as have previously concluded commercial treaties or trade agreements containing the most-favored-nation clause with one of the contracting parties.

Communist Chinese writer Wang Yih-wang defined the most-favored-nation clause in similar terms: "The meaning of the most-favored-nation clause is that when one of the contracting parties (for instance state A) grants certain favorable treatment to a third party, the other contracting party (state B) will automatically enjoy the same privilege."[76]

According to international practice, the most-favored-nation clause is generally applicable to commercial or economic matters; it is not meant for political or other matters. However, unilateral most-favored-nation clauses contained in the unequal treaties imposed upon China by Western states in the nineteenth century freely extended the application of this clause to political and other noncommercial matters. For instance, Article 30 of the Sino-American Treaty concerning Peace, Friendship and Commerce, signed at Tientsin on June 18, 1858, provided: "The Contracting Parties hereby agree that should at any time the Ta-Tsing [China] Empire grant to any nation, or the merchants or citizens of any nation, any right, privilege, or favor connected, either with navigation, commerce, *political or other intercourse*, which is not conferred by this Treaty, such

74. *LNTS*, 94:65.

75. Oppenheim, *International Law*, I, 971–972.

76. Wang Yih-wang, "What Is the Difference between Commercial Treaties and Trade Agreements," *Enlightenment Daily*, May 12, 1950, p. 3. Wang Yao-t'ien gives a similar definition in his *Kuo-chi mao-yi t'iao-yüeh ho hsieh-ting* (International trade treaties and agreements) (Peking, 1958), p. 25.

right, privilege, and favor shall at once freely enure to the benefit of the United States, its public officers, merchants, and citizens." (Emphasis added.) [77] This treaty has been severely criticized by Wang Yao-t'ien and is considered strong proof of American oppression of China in the past.[78]

Wang Yao-t'ien classifies the different forms of the most-favored-nation clause into five categories:

> (1) Conditional or unconditional – the conditional form provides that concessions shall be granted to a contracting state only upon the reciprocal payment of compensation equivalent to that paid by a third state for the concession; the unconditional form lays down no conditions for granting concessions to the contracting state.
>
> (2) Unilateral or mutual – the unilateral form grants one contracting state most-favored-nation treatment while denying it to the other; the mutual form consists of the reciprocal grant of most-favored-nation treatment between contracting states.
>
> (3) Unlimited or limited – the unlimited form imposes no restrictions on the scope of the application of the most-favored-nation clause; the limited form confines the application of the clause to certain specified objects or territories.
>
> (4) Positive or negative – the positive form requires either contracting state to undertake to grant the other all privileges, favors, and immunities it has granted or may hereafter grant to any other state; the negative form provides that neither contracting state shall treat the other less favorably than it does third states.
>
> (5) Simple or complex – the simple form is one which contains a general statement providing most-favored-nation treatment; the complex form defines the right in greater detail and usually consists of four parts dealing with general purposes, interpretation, limitations, and exceptions.[79]

Communist China has favored the unconditional, mutual, and limited forms of most-favored-nation clauses.[80] Wang Yao-t'ien states that the conditional form is "becoming rarely used nowadays" and that the uni-

77. Hertslet, *Hertslet's China Treaties* (London, 1908), I, pp. 551–552.

78. Wang Yao-t'ien, *International Trade Treaties*, pp. 30–31.

79. According to Gene T. Hsiao's study, Wang Yao-t'ien's classification scheme is based on a 1948 book by Richard Carlton Snyder (*The Most-Favored-Nation Clause* [New York, 1948], pp. 20–21, 52, 58). Snyder's work in turn drew heavily on an article by Stanley K. Hornbeck ("The Most-Favored-Nation Clause," *AJIL*, 3:395, 619 (1909); and Hornbeck, *The Most-Favored-Nation Clause* [Ann Arbor: University of Michigan Press, 1910]). Snyder and Hornbeck are cited in Hsiao, "Communist China's Trade Treaties and Agreements," *Vanderbilt L.R.*, 21:647n. 120 (October 1968). Description of the forms of the most-favored-nation clause is substantially based on Hsiao's study.

80. Hsiao, "Communist China's Trade Treaties," p. 647.

lateral form is "unequal."[81] A most-favored-nation clause is usually contained in Communist Chinese treaties of commerce and navigation and trade agreements. Generally, the right is not applied by Communist China to favors granted in frontier trade between neighboring states, or to trade between members of a customs union or any preferential system.

The following are two examples of Communist Chinese treaties containing most-favored-nation clauses.[82]

(1) Sino-Ceylonese Trade and Payment Agreement, September 19, 1957.[83]

Article VIII. The two Contracting Parties will grant to each other most-favored-nation treatment in respect of the issue of import and export licenses, and the levy of custom duties, taxes, and any other charges imposed on or in connection with the importation, exportation and trans-shipment of commodities, subject to the following exceptions: (1) Any special advantages which are accorded or may be accorded in the future by either of the Contracting Parties to contiguous countries in order to facilitate frontier trade and, (2) any special advantages which are accorded or may be accorded in the future under any preferential system of which either of the Contracting Parties is or may become a member.

(2) Sino-Soviet Treaty of Commerce and Navigation, April 23, 1958.[84]

Article 2. The Contracting Parties shall grant each other most-favored-nation treatment in all matters relating to trade, navigation and other economic relations between the two States.

Article 15. The provisions of this Treaty shall not extend to rights and advantages which may have been or may hereafter be granted by either of the Contracting Parties for the purpose of facilitating frontier trade relations with adjacent States in border areas.[85]

81. Wang Yao-t'ien, *International Trade Treaties*, pp. 29–31. But the mutual most-favored-nation clause does not necessarily mean that a treaty containing such a clause is equal. See *ibid.*, also chap. iv below, n. 41 and accompanying text.

82. For a survey of most-favored-nation clauses in various Communist Chinese treaties, see Hsiao, "Communist China's Trade Treaties," pp. 646–651.

83. *TYC*, 6:203; *UNTS*, 337:137.

84. *TYC*, 7:42; *UNTS*, 313:135.

85. A similar restrictive provision was contained in treaties of commerce and navigation with East Germany (*TYC*, 9:134), Albania (*TYC*, 10:290); and North Korea (*TYC*, 11:92). The Treaty of Commerce and Navigation with Outer Mongolia does not contain such a restrictive provision (*TYC*, 10:361).

IV Objects of Treaties

The objects of treaties can be obligations concerning any matter of interest to the contracting parties. However, international law does restrict or prohibit certain obligations from becoming the objects of treaties. Generally speaking, it imposes certain duties upon contracting parties:

(1) The duty not to conclude treaties inconsistent with the obligations of former treaties

(2) The duty not to conclude treaties inconsistent with the United Nations Charter

(3) The duty not to conclude treaties inconsistent with universally recognized principles of international law (*jus cogens*).

According to Lauterpacht's Oppenheim, the conclusion of treaties inconsistent with obligations of former treaties is "an illegal act which cannot produce legal results beneficial to the law-breaker." Nevertheless, this principle, in his view, should not be strictly applied to the modification of a general convention by a new treaty. In that case, "if it could be shown that the interests of the complaining state are not affected at all or that the degree to which they are affected is slight when related to the general advantage accruing from a new treaty," then the second treaty should not be held to be invalid. He further points out that this principle "does not apply to subsequent multilateral treaties, such as the Charter of the United Nations, partaking of a degree of generality which imparts to them the character of legislative enactments properly affecting all members of the international community or to such multilateral treaties as must be deemed to have been concluded in the general interest."[1]

Some Western writers do not agree with Lauterpacht's contention that, in general, treaties inconsistent with obligations of former treaties are invalid.[2] H. W. Briggs observes that this problem is "a matter of doctrinal dispute."[3] The 1969 Convention on the Law of Treaties does not adopt

1. Oppenheim, *International Law*, I, 8th ed. (London, 1955), 894–895.

2. In art. 16 of Lauterpacht's Second Draft on the Law of Treaties, prepared for the U.N. International Law Commission, he proposed the following rules governing treaties inconsistent with earlier treaties:

"1. A bilateral or multilateral treaty, or any provision of a treaty, is void if its performance involves a breach of a treaty obligation, previously undertaken by one or more of the contracting parties . . . 3. The above provisions apply only if the departure from the terms of the prior treaty is such as to interfere seriously with the interests of the other parties to that treaty or substantially to impair an essential aspect of its original purpose," *Yearbook of the International Law Commission, 1954*, 2:123, U.N. Doc. A/CN. 4/ser. A/1954/Add. 1.

3. Briggs, *Law of Nations* (New York, 1952), p. 848; yet he concluded that "the practice of states and the jurisprudence of national and international courts are in accord as to the legal priority of the earlier treaty." Parry wrote that Lauterpacht's

Lauterpacht's contention. Article 30, paragraph 4 (b), of the Convention provides: "as between a State party to both treaties and a State party to only one of the treaties, the treaty to which both States are parties governs their mutual rights and obligations."[4] But when the later treaty affects the interests of the state party to the earlier treaty, the Convention implies in Article 30, paragraph 5, that the later treaty does not lose its validity, but that the state party to the earlier treaty may invoke its right to terminate or suspend the operation of the earlier treaty under Article 60 of the Convention; or it may invoke the international responsibility of the party which has infringed its right.[5]

Communist China's practice on this problem has varied. In the following case concerning the 1951 Japanese Peace Treaty, its position seems to follow Lauterpacht's contention. On July 12, 1951, the United States government published its draft peace treaty with Japan, and on July 20 of the same year it invited all states at war with Japan during the Second World War, with the exception of China, to attend a conference in San Francisco the following September for the signing of a peace treaty with Japan. Communist China considered the United States' proposal inconsistent with earlier treaty obligations undertaken by the United States. On August 15, 1951, Foreign Minister Chou En-lai made a statement denouncing the proposal. The pertinent part of the statement follows:

> Whether considered from the procedure through which it was prepared or from its contents, the United States-British Draft Peace Treaty with Japan flagrantly violates those important international agreements to which the United States and British Governments were signatories, viz, the United Nations Declaration of January 1, 1942, the Cairo Declaration, the Yalta Agreement, the Potsdam Declaration and Agreement, and the Basic Post-Surrender Policy for Japan which was adopted by the Far Eastern Commission of June 19, 1947. The United Nations

---

view "is highly controversial . . . The better view would seem to be that the case will merely give rise to a right to damages if performance is not forthcoming, provided that the party entitled to performance contracted in good faith and in ignorance of the earlier promise" in "The Law of Treaties," in Sørensen, ed., *Manual of Public International Law* (New York, 1968), p. 207. See also Schwarzenberger, *International Law*, I (London, 1957), 275–276; McNair, *Law of Treaties* (London, 1961), pp. 218–224.

4. *ILM*, 8:691, no. 4 (July 1969).

5. *Ibid*. See also par. 11 of Commentary to art. 26 of Draft Articles on the Law of Treaties prepared by the U.N. International Law Commission (subsequently art. 30 of 1969 Convention on the Law of Treaties); Report of the International Law Commission on the 2nd Part of Its 17th Session (May 4–July 19, 1966), *AJIL*, 61:347 (1967).

Declaration provides that no separate peace should be made. The Potsdam Agreement states that the "preparatory work of the peace settlements" should be undertaken by those states which were signatories to the terms of surrender imposed upon the enemy state concerned . . .

The United States has monopolized the task of preparing the Draft Peace Treaty with Japan as now proposed, excluding most of the states that had fought against Japan and particularly the two principal Powers in the war, China and the Soviet Union, from the preparatory work for the peace treaty . . .

In violation of the agreement under the Cairo Declaration, the Yalta Agreement and the Potsdam Declaration, the Draft Treaty only provides that Japan should renounce all right to Taiwan and the Pescadores . . . without mentioning even one word about the agreement that Taiwan and the Pescadores be returned to the People's Republic of China . . .

With a view to expediting the concluding of a separate peace treaty with Japan, the United States Government, in its notification for the convocation of the San Francisco Conference, openly excludes the People's Republic of China – the principal Power which had fought against Japan – and thus completely violates a stipulation in the United Nations Declaration of January 1, 1942, to the effect that each of the signatory Powers pledged itself not to make a separate peace[6] . . .

Now, the Central People's Government of the People's Republic of China once again declares: If there is no participation of the People's Republic of China in the preparation, drafting and signing of a peace treaty with Japan, whatever the contents and results of such a treaty, the Central People's Government considers it all illegal, and therefore null and void.[7]

On September 8, 1951, the peace treaty was signed at San Francisco; on September 18 Chou En-lai again made a statement declaring that the peace treaty was "illegal and invalid."[8]

Communist Chinese writer Ch'en T'i-ch'iang has relied on Lauterpacht's

6. Declaration is listed in official U.S. treaty collections; EAS, no. 236, 55 Stat. 1600. It has also been continuously listed in Department of State's *Treaties in Force*. Communist China has always considered the declaration a treaty; it is listed in *Kuo-chi t'iao-yüeh chi* (International treaty series), 1934–1944 (Peking, 1961), 3:342.

7. *WCC*, 2:30–36; Egnlish trans. in "Foreign Minister Chou En-lai's Statement on U.S.-British Draft Peace Treaty with Japan and San Francisco Conference," *NCNA*, Daily News Release no. 777 (Aug. 16, 1951), pp. 75–78.

8. *WCC*, 2:37. The peace treaty came into force Apr. 28, 1952. On May 5, 1952, Chou En-lai made another statement declaring that it was "completely illegal and unreasonable," *WCC*, 2:66. In both the Sept. 18, 1951, and May 5, 1952, statements, Chou En-lai repeatedly emphasized that the peace treaty was in violation of former treaties agreed upon between allies.

Oppenheim to support the view that the above-mentioned Japanese peace treaty is "illegal and, therefore, has no legal force."[9] Nevertheless, Communist China has not always maintained that no treaty should be concluded in violation of former treaty obligations. When in 1958 the Soviet Union proposed to change the status of Berlin that had been prescribed by the 1945 Yalta and Potsdam agreements, Communist China supported the Soviet proposal – though it argued that those agreements had been violated by the United States, the United Kingdom, and France, which had thereby forfeited their legal basis to occupy West Berlin.[10]

Moreover, not all Communist Chinese writers insist on the invalidity of a treaty which is inconsistent with a former treaty. For example, when the 1954 Paris Agreements[11] were signed by the United Kingdom, France, West Germany, and the United States, the Soviet Union abrogated the Treaties of Mutual Assistance concluded with France and the United Kingdom during the Second World War. A writer in the *People's Daily* of April 12, 1955, supported the Soviet position as "perfectly proper and necessary," reasoning that the Paris Agreements were inconsistent with the obligations of the treaties abrogated by the Soviet Union.[12] The article did not claim that the Paris Agreements were invalid

9. He writes: "The San Francisco [peace] treaty [concluded with Japan in 1951], as everybody knows, was signed in violation of the terms of [the United] Nations Declaration of January 1, 1942, prohibiting the making of a separate peace with the enemy and also in violation of the obligation under the Cairo Declaration to restore Taiwan and the Penghu Islands to China. The 'treaty' is illegal and, therefore, has no legal force.

"It is clearly stated in Oppenheim's *International Law* that a state has 'the duty not to conclude treaties inconsistent with the obligations of former treaties. The conclusion of such treaties is an illegal act which cannot produce legal results beneficial to the law breaker,'" in "Sovereignty of Taiwan belongs to China," *People's Daily*, Feb. 9, 1955, p. 4; English trans. entitled "Taiwan--A Chinese Territory," *Law in the Service of Peace*, (Brussels, 1956), V, 42–43.

10. See "The Declaration of the Government of the People's Republic of China Supporting the Plan of the Soviet Union to Withdraw Foreign Forces from Berlin and Terminate the Berlin Occupation," Dec. 21, 1958; *WCC*, 5:219–220.

11. The Paris Agreements were concluded Oct. 23, 1954; they consist of (1) Protocol to North Atlantic Treaty on Accession of the Federal Republic of Germany; (2) Protocol on Termination of the Occupation Regime in the Federal Republic of Germany; (3) Convention on the Presence of Foreign Forces in Germany; *DIA* (1954), pp. 102–107.

12. Tan Wen-jui, "Don't Allow the Use of International Treaties as a Smoke Screen," p. 4. On Mar. 10, 1923, the Republic of China wrote to the Japanese Government requesting abrogation of the "Twenty-one Demands" included in the notes exchanged between China and Japan on May 25, 1915. One of the reasons invoked by China to support its request was that the demands "are in violation of the treaties between China and the other powers," Hackworth, *Digest of International Law* (Washington, 1943), 5:162.

because of this inconsistency, but it asserted incompatibility to justify the Soviet abrogation.

Article 103 of the United Nations Charter provides: "In the event of a conflict between the obligations of the Members of the United Nations under the present Charter and their obligations under any other international agreement, their obligations under the present Charter shall prevail." According to Lauterpacht's Oppenheim, Article 103 makes treaties inconsistent with the Charter "for all practical purposes, void and unenforceable."[13] This view is not shared by some Western jurists,[14] though Soviet jurists[15] and Communist Chinese writers (discussed later) appear to accept Lauterpacht's construction. The 1969 Convention on the Law of Treaties does not go so far as Lauterpacht does on this question. Article 52 of the Convention says: "A treaty is void if its conclusion has been procured by the threat or use of force in violation of the principles of international law embodied in the Charter of the United Nations."[16] Even assuming the correctness of Lauterpacht's explanation, its application to specific cases will inevitably be controversial.

The United Nations Charter does not state which organ is competent to render authoritative interpretations of the Charter's provisions. According to the report of the 1945 San Francisco Conference: "The members

13. Oppenheim, *International Law*, I, 896.

14. Kelsen's interpretation of art. 103 of the Charter is similar to Lauterpacht's; see his *The Law of the United Nations* (London, 1950), p. 113. But Parry observes that art. 103 does not state "in terms that an inconsistent later treaty is to be regarded void [and] that in the Charter perhaps suggests more strongly that what may require consideration is not only a possible collision between the actual letter of the Charter and a subsequent treaty but also between a specific duty arising out of action taken under the Charter and the latter . . . [therefore] it is difficult to regard Article 103 as implying the illegality, as opposed to the mere liability to suppression, of any category of subsequent treaties," in Sørensen, *Manual*, p. 208. O'Connell also feels that art. 103 "carefully avoids stating that the inconsistent treaty is invalid, and it may be that its only effect is to prevent members from invoking the inconsistent treaty before the United Nations organ," *International Law*, I, (London, 1965), 293. Similarly, Goodrich and Hambro wrote: "It is to be noted that this Article does not provide for the automatic abrogation of obligations inconsistent with the terms of the Charter. The rule is put in such form as to be operative only when there is an actual conflict," *Charter of the United Nations: Commentary and Documents* (London, 1949), p. 519; see also the Goodrich, Hambro, and Simons 3rd and rev. ed. (1969), p. 615, where it is observed that at the 1945 San Francisco Conference "it was thought inadvisable to provide for the automatic abrogation by the Charter of obligations inconsistent with the terms thereof."

15. See Triska and Slusser, *The Theory, Law, and Policy of Soviet Treaties* (Stanford, 1962), pp. 119–121.

16. *ILM*, 8:698, no. 4 (July 1969).

or the organs of the organization might have recourse to various expedients [such as an ad hoc commission of jurists or an advisory opinion of the International Court of Justice] in order to obtain an appropriate interpretation . . . It is to be understood . . . that, if an interpretation made . . . is not generally acceptable, it will be without binding force."[17] Consequently, the question whether a treaty is inconsistent with the United Nations Charter is always debatable. It is not surprising that such broad terms as "sovereign equality," "domestic jurisdiction," "human rights," and "collective self-defense," have been given divergent interpretations by members of the United Nations.

The following two cases illustrate the types of treaties which Communist Chinese scholars have considered inconsistent with the United Nations Charter. On December 2, 1954, the Republic of China and the United States signed a Treaty of Mutual Defense[18] that in the view of those states falls within the meaning of the collective self-defense provision of Article 51 of the Charter. Communist Chinese writer Shao Chin-fu, however, views the treaty in a very different light:

> Article 103 of the U.N. Charter states: "In the event of a conflict between the obligations of the Members of the United Nations under the present Charter and their obligations under any other international agreement, their obligations under the present Charter shall prevail." In other words " . . . to the extent of their [the treaties'] inconsistency with the Charter all these agreements are, for all practical purposes, void and unenforceable."[19] The aim of the U.S.-Chiang "treaty" is "to resist armed attack and communist subversive activities directed from without against their territorial integrity and political stability." (Article II). Hence, "to resist armed attack" means to resist the Chinese people's liberation of Taiwan, on the pretext of which the United States can provoke war at any time. To resist "communist subversive activities" means that the United States can send troops to suppress the Taiwan people, because the United States and the Chiang Kai-shek clique are always in the habit of calling any revolutionary activity as "directed from without." The contents of the U.S.-Chiang "treaty" are to empower the United States to split China, provoke war and interfere in China's internal affairs. This, of course, contravenes the

17. Doc. no. 933, IV/2/42(2), *United Nations Conference on International Organization Documents* (1945), XIII, 710.
18. *UST*, 6:433; *TIAS* 3178; *UNTS*, 248:213.
19. Shao cites this passage from Oppenheim, *International Law*, I, 7th ed. (London, 1948), 807.

> principles of international law and the U.N. Charter, and what contravenes these principles has, of course, no validity at all.[20]

Similarly, Professor Ch'iu Jih-ch'ing of Fu-tan University at Shanghai argues that the North Atlantic Treaty, Baghdad Treaty, and Southeast Asia Collective Defense Treaty are in violation of the United Nations Charter and are therefore "void."[21]

According to Lauterpacht's Oppenheim, obligations which are inconsistent with universally recognized principles of international law cannot be the object of a treaty.[22] This view is generally accepted by jurists, though there is no consensus on what constitutes "recognized principles of international law."

The 1969 Convention on the Law of Treaties also fails to make this point clear. Article 53 of the Convention provides: "A treaty is void if, at the time of its conclusion, it conflicts with a peremptory norm of general international law [*jus cogens*]. For the purposes of the present Convention, a peremptory norm of general international law is a norm accepted and recognized by the international community of States as a whole as a norm from which no derogation is permitted and which can be modified only by a subsequent norm of general international law having the same character."[23] The commentary to this Article, prepared by the United Nations International Law Commission, points out that "the formulation of this article is not free from difficulty, since there is no simple criterion by which to identify a general rule of international law as having the character of *jus cogens*."[24]

The question becomes even more complicated when one attempts to discover what kind of principles Communist China considers to be "universally recognized principles of international law." One important area

20. Shao Chin-fu, "The Absurd Theory of 'Two Chinas' and Principles of International Law," *KCWTYC*, 2:14 (1959); English trans. in *Oppose the New U.S. Plots to Create "Two Chinas"* (Peking, 1962), pp. 89–90. For a criticism of this treaty, see Ch'in Tzu-ch'iang, *Lun Mei-Chiang ch'in-lüeh t'iao-yüeh* (On the American-Chiang Kai-shek treaty of aggression) (Peking, 1955).

21. Ch'iu Jih-ch'ing, "Further Discussion of the System of International Law at the Present Stage," *FH*, 3:43 (1958).

22. Oppenheim, *International Law*, I, 897.

23. *ILM*, 8:698–699, no. 4 (July 1969).

24. Par. 2 of Commentary to art. 50 (subsequently art. 53 of 1969 Convention on the Law of Treaties); *AJIL*, 61:410 (1967).

of this type of disagreement is the concept of unequal treaties. This concept is regarded by all Communist scholars as a very fundamental principle of international law, but it is not so considered by most contemporary Western scholars – a problem that merits expansion.

Although the concept of unequal treaties has been championed by Communist scholars, one must realize that it is neither a new nor a Communist invention. Several classical Western writers such as Grotius, Pufendorf, Vattel, and Wolff have examined this problem. Grotius, for instance, in his famous *De Jure Belli Ac Pacis* published in 1646, wrote that treaties "which add something beyond the rights based on the law of nature are either on equal or on unequal terms."[25] None of these writers challenged the legal validity of unequal treaties, though Vattel did argue that "since Nations are no less bound than individuals to respect justice, they should make their treaties equal, as far as that is possible."[26]

In the nineteenth century, Western interest in the concept of unequal treaties began to decline; since the beginning of this century, the general tendency of Western treatises has been to ignore the problem.[27] After the First World War, the issue of unequal treaties received widespread attention when Soviet Russia offered to abolish many of the treaties formerly imposed upon some Asian states by Czarist Russia.[28] At the same time, China demanded the abolition of *some* treaties which it termed "unequal." The term "unequal treaties" as used in that period generally referred to those nineteenth- and early twentieth-century treaties by which the Western powers, including Czarist Russia, forced China and other Asian states to accept consular jurisdiction (extraterritoriality), unilateral most-favored-nation treatment, restrictive tariff regulations, territorial cessions or leases, and other disabilities.

When the Nationalist party assumed power in China in 1928, it proclaimed its determination to abolish or revise unequal treaties through diplomatic negotiations.[29] The Nationalist government experienced some

25. Grotius, *De jure belli ac pacis*, II, trans. Kelsey (Washington, 1925), 394.

26. Vattel, *Le Droit de gens: ou principes de la loi naturelle*, III, trans. Fenwick (Washington, 1916), 165.

27. The term is not found, in connection with discussion of treaties, in treaties of international law written by such nineteenth- and twentieth-century Western scholars as Wheaton, Woolsey, Phillimore, Brierly, Oppenheim, Lauterpacht, Hyde, Briggs, Fenwick, Hall, Jessup, Kelsen, and O'Connell; only recently have a few Western writers begun to include brief discussion of unequal treaties, *e.g.*, Brownlie, *Principles of Public International Law* (London, 1966), pp. 495–496.

28. *E.g.*, see the 1919 and 1920 Soviet Declarations to China. Woodhead, *China Year Book 1924–25* (Tientsin, n.d.), pp. 868–872.

29. See the Nationalist Government Declaration of June 16, 1928; *Chinese Social and Political Science Review, Public Documents* (Peking, 1928), 12:47–48. At that

success relative to treaties imposing restrictive tariff provisions and allowing foreign settlement in several Chinese cities. But no substantial progress was made in the area of consular jurisdiction.

It was for this reason that after three years of fruitless negotiation with the Western powers, the Nationalist government decided unilaterally to abolish consular jurisdiction in China. On May 4, 1931, the Nationalist government promulgated a law which was to come into force on January 1, 1932.[30] But on September 18, 1931, Japan suddenly occupied Mukden; soon afterwards it invaded all Manchuria. Preoccupied by the Manchurian situation, the Nationalist government issued a decree on December 29, 1931, indefinitely postponing enforcement of the May 4 law.[31]

On January 11, 1943, the United States and the United Kingdom concluded new treaties with China which abolished extraterritoriality and other rights in China.[32] The remaining extraterritorial and other rights in China of other states were all liquidated soon after the conclusion of the Second World War.[33]

After 1943 the problem of unequal treaties received little attention from Nationalist leaders and scholars (two exceptions will be mentioned later). The Communist Chinese, however, persisted in their interest in the subject. Though more vocal than the Nationalist Chinese on the question of "unequal treaties," the Communist Chinese have never clearly defined what they mean by this term. A textbook on international trade treaties, used by the Peking Foreign Trade Institute, states:

> The classical writers of Marxism-Leninism confirmed an important principle concerning international treaties, namely, the genuine sovereign equality between all parties concerned should become the foundation of international treaties. Lenin said: Negotiation can only be conducted between equals and, therefore, genuine equality between both sides is an essential condition for reaching a genuine agreement.
>
> Consequently, in accordance with Marxism-Leninism, there are equal treaties and unequal treaties, and, therefore, progressive mankind takes

---

time the Nationalist effort to abolish China's unequal treaties did not include treaty provisions concerning territorial cessions, such as the provisions of the 1842 Nanking Treaty ceding Hong Kong to Britain, or the 1860 Peking Treaty ceding Chinese territory east of the Ussuri River to Czarist Russia. For a study of the Nationalist effort to abolish unequal treaties, see Tseng Yu-hao, *Termination of Unequal Treaties in International Law* (Shanghai, 1933); Fishel, *End of Extraterritoriality in China* (Berkeley and Los Angeles, 1952).

30. "Regulations Governing the Exercise of Jurisdiction over Foreign Nationals in China," in Woodhead, *China Yearbook 1932* (Shanghai, n.d.), p. 263.

31. *Ibid.*, p. 264.

32. 57 Stat. 767; *TS* no. 984; *UNTS*, 10:261 (U.S.); *LNTS*, 205:73 (U.K.).

33. See Fishel, *End of Extraterritoriality*, pp. 214–215.

> fundamentally different attitudes towards different kinds of treaties. Equal treaties should be strictly observed. Unequal treaties are in violation of international law and without legal validity.[34]

Similarly, a 1958 Chinese article reasons: "Treaties can be classified into equal treaties and unequal treaties, and the latter undermine the most fundamental principles of international law – such as the principle of sovereignty; therefore, they are illegal and void, and states have the right to abrogate this type of treaty at any time."[35]

These illustrative opinions conform generally to the position of most Soviet scholars. Kozhevnikov, for example, states: "Equal treaties are treaties concluded on the basis of equality between the parties; unequal treaties are those which do not fulfill this elementary requirement [and] . . . are not legally binding."[36]

Recently, the 1969 United Nations Conference on the Law of Treaties, in its Final Act, adopted a declaration on "the Prohibition of Military, Political or Economic Coercion in the Conclusion of Treaties." The declaration "solemnly condemns the threat or use of pressure in any form, whether military, political or economic, by any State in order to coerce another State to perform any act relating to the conclusion of a treaty in violation of the principles of sovereign equality of States and freedom of consent."[37] Thus in principle the Communist Chinese and the Soviet position does not seem to differ from that of current Western or non-Communist states.

But the time element remains for applying the principles contained in the declaration. This declaration says that "the United Nations Conference on the Law of Treaties . . . *deplor[es]* the fact that in the past States have sometimes been forced to conclude treaties under pressure exerted in various forms by other States [and d]esir[es] to ensure that in the future no such pressure will be exerted in any form by any State in connexion with the conclusion of a treaty." Apparently this declaration is designed to announce principles for future conduct of states. Communist Chinese and Soviet theory, however, suggests that these two countries consider that the principles contained in the declaration have

34. Wang Yao-t'ien, *Kuo-chi mao-yi t'iao-yüeh ho hsieh-ting* (International trade treaties and agreements) (Peking, 1958), p. 10.
35. Shih Sung, *et al.*, "An Initial Investigation into the Old Law Viewpoint in the Teaching of International Law," *CHYYC*, 4:14 (1958).
36. Kozhevnikov, ed., *International Law* (Moscow [1961]), p. 248.
37. *ILM*, 8:733, no. 4 (July 1969).

long been rules of international law and are applicable to all treaties, even those concluded in the past.

Although both Soviet and Communist Chinese scholars consider "unequal treaties" to lack legally binding force, what they mean by that term remains unclear. Communist China, like the Chinese Nationalist government, considers all treaties imposed on China in the nineteenth and early twentieth centuries (that is, those concerning consular jurisdiction, unilateral most-favored-nation treatment, restrictive tariff regulations, territorial cessions or leases, and other disabilities) to be unequal treaties.[38]

Communist Chinese scholars maintain that verbal reciprocity alone does not make a treaty "equal" if important political and economic facts are not taken into consideration. Wang Yao-t'ien has written that "whether or not a treaty is equal does not depend upon the form and words of various treaty provisions, but depends upon the state character, economic strength, and the substance of correlation of the contracting states."[39] In this respect, several Communist Chinese writers consider the 1946 Sino-American Treaty of Friendship, Commerce, and Navigation,[40] concluded by the Nationalist government, to be an unequal treaty despite

38. *E.g.*, see "A Comment on the Statement of the Communist Party of the U.S.A." (editorial), *People's Daily*, Mar. 8, 1963, p. 1. Before assuming power the Chinese Communist party also advocated abolition of China's unequal treaties; at that time its concept of unequal treaties was similar to that of the Nationalist party. A statement on the current situation by the Central Executive Committee of the Chinese Communist Party on June 15, 1922, stated that one party goal was to "rectify [China's] conventional tariff system, to cancel various extraterritorial rights of major powers in China, and to regain the administration of the Chinese railway by way of repaying all railway loans," condensed text of statement in *Chung-kuo wen-t'i chih-nan* (Guide on China's problems) (Yenan, 1937), II, 19–31; reprinted in *Kung-fei huo-kuo shih-liao hui-pien* (Collection of historical materials relating to the Communist bandits' rebellion) (Taipei, 1964), I, 17. A declaration issued by the same organ on July 10, 1925, called for support of the Nationalist party's efforts "to declare the abrogation of all unequal treaties," *ibid.*, pp. 64–69, reprint p. 45.

A declaration issued by the Central Committee of the Chinese Communist Party on July 13, 1927, proclaimed that "the Chinese Communists . . . shall make efforts to abolish all unequal treaties, to take back all settlements, to cancel extraterritoriality, to practice tariff autonomy, and to liberate China," *Kuo-wen chou-pao* (National news weekly), 4:29 (July 31, 1927); reprint *ibid.*, p. 347. In an outline of the struggle over the Shanghai incident (Japanese attack of Shanghai Jan. 28, 1932), the Chinese Communist party demanded "unilateral abrogation of all unequal treaties with China imposed by any imperialism," *Kung-fei huo-kuo shih-liao hui-pien*, 2:149. See also Mao Tse-tung, "The Chinese Revolution and the Chinese Communist Party," December 1939, in *Selected Works of Mao Tse-tung* (Peking, 1965), II, 311.

39. Wang Yao-t'ien, *International Trade Treaties*, p. 31.

40. 63 Stat. 1299; *TIAS* 1871; *UNTS*, 25:69.

the fact that the treaty granted, inter alia, reciprocal commercial privileges to each country. One Communist Chinese scholar has criticized the treaty thus:

> There was a drastic difference in economic strength between old China and American imperialism; old China was a backward country dependent upon the United States, and the Chinese bourgeoisie was extremely weak. How could the Chinese bourgeoisie have had the ability to establish factories in the United States, to carry on commerce, and to engage in scientific and cultural enterprises in order to realize the rights given to them by the "Sino-American Treaty of Friendship, Commerce and Navigation"? Only the American monopolistic bourgeoisie and its representatives were in a position really to enjoy these rights.[41]

Along such lines, Communist China's own record is not spotless. For instance, a charge of verbal reciprocity in disregard of important political facts could be aimed at the 1960 Sino-Burmese Treaty of Friendship and Mutual Non-Aggression.[42] Article 3 provides that "each Contracting Party undertakes . . . not to take part in any military alliance directed against the other Contracting Party." Communist China, a country with 800 million population and 3 million in its armed forces, does not need an ally to protect itself against Burma, a country with 14 million population and 100,000 in its armed forces. Thus, though equal on its face, the treaty could in fact only operate so as to leave Burma unprotected by prohibiting an alliance with the country which could most effectively resist Communist China's expansion, namely, the United States. In fact, an editorial in the *People's Daily* on February 1, 1960, makes it clear that the purpose of this treaty was, inter alia, to thwart United States policy in Southeast Asia.

41. Ch'ien Szu, "A Criticism of the Views of Bourgeois International Law on the Question of the Population," *KCWTYC*, 5:40, 47 (1960). Chou Keng-sheng has expressed the same view: "Even those treaties which provide mutual benefit in form but, due to the unequal economic position between the two contracting parties, in fact yield unilateral benefits to one side and subject the other side only to exploitation, are neither treaties of mutual benefit nor equal international transactions. The most obvious example is the so-called Sino-American Treaty of Commerce concluded in 1946 between American imperialists and the Kuomintang Government," in "The Principles of Peaceful Coexistence from the Viewpoint of International Law," *CFYC*, 6:41 (1955). Similarly, the U.S.-Italian Treaty of Friendship, Commerce, and Navigation (Feb. 2, 1948), *UNTS*, 79:171, was considered an unequal treaty. See Wang Yao-t'ien, *International Trade Treaties*, pp. 51–52.

42. *TYC*, 9:44; English trans. in "Treaty of Friendship and Mutual Non-Aggression between the People's Republic of China and the Union of Burma," *PR*, 3:13, no. 5 (Feb. 2, 1960).

The following treaties, considered unequal by the Communist Chinese, illustrate the range of this concept as applied by Peking:

(1) In 1956 the United States concluded an agreement with Switzerland on cooperation concerning civil uses of atomic energy.[43] According to the agreement, the United States would supply 500 kilograms of fissionable materials to Switzerland, as well as certain information and equipment. The United States, however, reserved the right to dispatch personnel to inspect or supervise the use of the materials; it was also granted the right to acquire or use Swiss inventions based on nuclear information supplied by the United States. This agreement was criticized by an article in the *People's Daily* of December 31, 1956, as a typical example of an unequal treaty concluded in the name of international cooperation.[44]

(2) Also in 1956 Jordan abolished the 1948 alliance treaty[45] with Britain which granted Britain two military bases in Jordan and the right to send troops into Jordan in time of war. A commentary in the *People's Daily* of November 29 by "Observer," the pseudonym for a senior Communist official, applauded the Jordanian action, observing that it abolished an unequal treaty.[46]

(3) In 1965, the United States and the Republic of China concluded an agreement on the status of the United States' forces in China which granted those forces the privileges of exemption from visas and taxes, immunity from local jurisdiction over acts committed in the course of "official duty," the right to establish post offices, and other privileges.[47] The *People's Daily* of September 13 strongly denounced the agreement as an unequal treaty.[48] It is interesting to note that some leaders of the Republic of China felt the same way. Ch'iao Yi-fan, a member of the Legislative Yüan, commented at length on the agreement, calling on the United States not to force the Nationalist Chinese government to grant these privileges; to do so would only provide propaganda material for

43. *UST*, 8:91; *TIAS* 3745; *UNTS*, 278:41.
44. Huang Yü, "Such Cooperation," p. 6.
45. *UNTS*, 77:77.
46. "Jordan Announces the Abrogation of the Anglo-Jordanian Treaty," p. 6.
47. *UST*, 7:373; *TIAS* 5989.
48. "U.S.-Chiang Illegally Signed 'Status Agreement' concerning U.S. Force of Aggression," *People's Daily*, p. 2; see also "U.S.-Chiang Kai-shek Illegal 'Status Agreement': A New Step to Make Taiwan U.S. War Base," *People's Daily*, Feb. 14, 1966, p. 2. English trans. in *SCMP*, 3640:26 (Feb. 17, 1966); *Commentator*, "Another Crime of the U.S.-Chaing Kai-shek Conspiracy," *People's Daily*, Feb. 19, 1966, p. 1, English trans. in "*Jen-min Jih-pao* Commentator Condemns U.S.-Chiang Kai-shek Illegal 'Status Agreement,'" *SCMP*, 3644:41 (Feb. 19, 1966).

Communist China in its attempt to undermine cooperation between the Republic of China and the United States.[49]

(4) On July 6, 1946, the United States and the Philippines concluded a trade agreement which provides, inter alia, ceilings for Philippine exports to the United States (Article 3); no restrictions are provided for United States exports to the Philippines.[50] It also (Article 5) restricts the freedom of the Philippines to change its currency value in relation to United States dollars and to alter the convertibility of Philippine currency into United States dollars. This agreement was denounced by a Communist writer as an unequal treaty.[51]

(5) On August 21, 1968, the armed forces of the Soviet Union, East Germany, Hungary, Poland, and Bulgaria suddenly invaded Czechoslovakia and soon occupied the whole country.[52] The occupation was denounced by a proclamation of the Presidium of the Czechoslovakian National Assembly issued on the same day.[53]

Under Soviet military occupation, Czechoslovakia signed on October 16, 1968, an agreement with the Soviet Union on "the conditions of temporary sojourn of Soviet forces on the territory of the Czechoslovak Socialist Republic."[54] On October 21, an *NCNA* news commentary claimed that the treaty

> unscrupulously tramples on Czechoslovak sovereignty and reduces the country to a virtual vassalage of Soviet revisionist social-imperialism . . .[It] does not define how long "temporary" means. Nor does it indicate when the "security" of what the Soviet revisionists call the "socialist community" no longer needs "defending." Instead, the traitorous treaty stipulates that it shall remain in force as long as the Soviet revisionists' troops are stationed on Czechoslovak territory. This

49. See Ch'iao Yi-fan's written opinion to the Executive Yüan, Jan. 4, 1966; *Li-fa-yüan kung-pao* (Gazette of the Legislative Yüan), 36th sess., 8:59 (Taipei, Jan. 14, 1966).

50. *UNTS*, 43:136.

51. C. K. Cheng, "The Philippines: America's Show-Window of Democracy in Asia?" *PR*, 8:21, no. 6 (Feb. 5, 1965). The article also listed the following U.S.-Philippine treaties as unequal treaties: (1) 1955 Laurel-Langley Agreement (revising the 1946 Trade Agreement), *UST*, 6:2981; *TIAS* 3348; *UNTS*, 238:264; (2) 1947 U.S.-Philippine Military Bases Agreements, 61 Stat. 4019; *TIAS* 1775; *UNTS*, 43: 271; (3) 1947 U.S.-Philippine Military Aid Agreement, 61 Stat. 3283; *TIAS* 1662; *UNTS*, 45:47; (4) 1951 U.S.-Philippine Mutual Defense Treaty, *UST*, 3:3947; *TIAS* 2529; *UNTS*, 177:133. See also Wang Yao-t'ien, *International Trade Treaties*, p. 50, condemning the 1946 U.S.-Philippine Trade Agreement as an unequal treaty.

52. See "Tass Statement on Soviet Military Intervention," Aug. 21, 1968, reprinted in *ILM*, 7:1283; no. 6 (November 1968).

53. Proclamation reproduced in *Ibid.*, 1286.

54. Trans. into English by W. E. Butler in *ibid.*, 1334–1339.

> means in fact that the military occupation of Czechoslovakia by the Soviet revisionists' troops will be indefinitely prolonged.[55]

The examples given above all concern bilateral treaties, but this does not mean that the concept of unequal treaties is inapplicable to multilateral treaties. An example of a multilateral treaty which is considered unequal by Communist China is the 1963 Partial Nuclear Test-Ban Treaty.[56] An article in the August 10, 1963, *People's Daily* by "Observer"[57] severely criticized the treaty because it places the three big nuclear powers – the United States, the United Kingdom, and the Soviet Union – in a privileged position and does not create identical and reciprocal obligations among all contracting parties. The article complains that other countries are induced to undertake an unconditional obligation not to test any nuclear weapons in the atmosphere while those three powers can continue underground testing. Because the testing of nuclear weapons must begin with atmospheric tests, it would be impossible for other states ever to manufacture nuclear weapons; thus they are relegated to a position of permanent inferiority. Moreover, the article comments that the provision for amendment in the treaty also gives the privilege of veto to each of the three big nuclear powers. Thus, no matter how many smaller states accede to the treaty, they are without the right or power to modify it if one of the three disagrees.

Another recent example of a multilateral treaty considered unequal by Communist China is the "Treaty of Non-Proliferation of Nuclear Weapons"[58] adopted by the United Nations General Assembly on June 12, 1968.[59] An article in the June 13 *People's Daily* by "Commentator," a pseudonym for a senior Communist Chinese official, denounced the treaty as a

> thoroughly unequal treaty dished up by the U.S. imperialists and Soviet revisionists . . . Under this treaty, the U.S. imperialists and Soviet revisionists are not only allowed to produce and stockpile nuclear weapons and increase the number of their nuclear bases; they

55. "Diabolical Social-Imperialist Face of the Soviet Revisionist Renegade Clique," *PR*, 11:8, no. 43 (Oct. 25, 1968). The *New York Times* also considered this an unequal treaty; see its editorial, "Prague's 'New Reality,'" Oct. 21, 1968, p. 46.

56. *UST*, 14:1313; *TIAS* 5433; *UNTS*, 480:43.

57. "Why the Tripartite Treaty Does Only Harm and Brings No Benefits," p. 1; English trans. in *PR*, 8:20, no. 33 (Aug. 16, 1963).

58. Text in *ILM*, 7:811–817, no. 4 (July 1968).

59. U.N. Doc. A/RES/2373 (XXII), June 18, 1968.

> also undertake no commitment whatsoever not to use nuclear weapons against the non-nuclear states. The latter, on the other hand, are totally deprived of their right to develop nuclear weapons for self-defense and are even restricted in their use of atomic energy for peaceful purposes. As the saying goes, "the magistrates are allowed to burn down houses while the common people are forbidden even to light lamps." In reality, this is tantamount to a demand that other countries accept forever the U.S. imperialist and Soviet revisionist position of nuclear monopoly and place themselves at their mercy.[60]

It must be pointed out that the principle suggested by the *People's Daily* of "identical and reciprocal" obligations among all contracting parties under a multilateral treaty is not always followed by Communist Chinese officials and scholars. For instance, the United Nations Charter accords special veto power to the Big Five in the Security Council and in regard to Charter amendment. Yet neither Communist China nor its scholars have ever maintained that these features of the Charter make it an "unequal treaty." On the contrary, Communist Chinese scholars have frequently attacked the alleged attempt of "American imperialism" to eliminate the Big Five's veto in the United Nations.[61]

Application of the concept of "unequal treaties" by Communist China is flexible and seems largely determined by political considerations, as the following cases illustrate.

(1) In 1945 the Republic of China, under the influence of the 1945 Yalta agreement concluded by the United States, the United Kingdom, and the Soviet Union, concluded a treaty of friendship and alliance and accompanying agreements[62] with the Soviet Union; these documents granted to the latter the right of joint administration and possession over

60. "A Nuclear Fraud Jointly Hatched by the United States and the Soviet Union," p. 5; English trans. in *PR*, 11:17, no. 25 (June 21, 1968). The treaty was denounced by Tanzanian President Nyerere as "a most unequal treaty," see his speech at Peking, June 21, 1968, "President Nyerere's Speech," *PR*, 11:7, no. 26 (June 28, 1968). The pro-Peking Afro-Asian Writers' Bureau issued a statement on July 4, 1968, denouncing this "unequal treaty"; "Afro-Asian Writers' Bureau Denounces 'Treaty on Non-Proliferation of Nuclear Weapons,'" *PR*, 11:6, no. 29 (July 19, 1968).

61. *E.g.*, see Kuo Ch'un, *Lien-ho kuo* (The United Nations) (Peking, 1956), pp. 101–107.

62. Sino-Soviet Agreement on Chinese Ch'ang Ch'un [Chinese Eastern] Railway (Aug. 14, 1945), *UNTS*, 10:346; Sino-Soviet Agreement on Port Arthur (Aug. 14, 1945), *ibid.*, 358; Sino-Soviet Agreement on Dairen (Aug. 14, 1945), *ibid.*, 354; Exchange of Notes on the Independence of Outer Mongolia (Aug. 14, 1945), *ibid.*, 344.

the Chinese Eastern Railway, a military base in Port Arthur, special privileges in the port of Dairen, and many other privileges. At the request of the Soviet Union, the Republic of China also recognized the independence of Outer Mongolia, thus relinquishing the long-held Chinese claim to sovereignty over Mongolia. By definition, the agreements would seem to be unequal treaties because the Soviet Union did not grant to China any reciprocal privileges in respect to Soviet territory. Nevertheless, neither Communist Chinese leaders nor scholars have ever pubicly denounced these agreements, though the record of Soviet-Communist Chinese relations amply demonstrates that Communist Chinese leaders are not happy with them. In 1950 the Soviet Union agreed to hand over to Communist China all Soviet rights over the Chinese Eastern Railway not later than 1952, to evacuate its Port Arthur military base, and to renounce its privileges in the port of Dairen.[63] As for Outer Mongolia, Communist China in a 1950 exchange of notes with the Soviet Union recognized Mongolia's independent status;[64] but in 1954 it reopened this issue during Khrushchev's visit to China.[65]

(2) Until recently, Communist Chinese leaders and scholars had been publicly silent on the postwar treaties by which the Soviet Union acquired a large portion of territory from Eastern European countries.[66] With the intensification of the Sino-Soviet dispute, however, Communist China has begun to attack the Soviet Union for those treaties. In an interview with a Japanese Socialist delegation in 1964, Chairman Mao severely criticized postwar Soviet annexation of certain territories of Eastern Europe, thus implying that those treaties of annexation are unequal and, consequently, subject to renegotiation.[67] Mao's statement was sharply repudiated by an editorial in *Pravda* which charged that Mao "crossed out with amazing ease the entire system of international agreements concluded after World War II which meet the interests of strength-

63. Sino-Soviet Agreement concerning the Chinese Ch'ang Ch'un Railway, Port Arthur and Dairen (Feb. 14, 1950), *UNTS*, 226:31.

64. *Ibid.*, 16.

65. "On Mao Tse-tung's Talk with a Group of Japanese Socialists" (editorial), *Pravda*, Sept. 2, 1964; English trans. in *International Affairs*, 10:80 (Moscow, 1964).

66. Mao Tse-tung supported 1939 Soviet occupation of the eastern part of Poland; "The Identity of Interests between the Soviet Union and All Mankind," *Selected Works of Mao Tse-tung*, II, English version (Peking, 1965), p. 279. The Soviet-occupied Polish territory was formally annexed by the Soviet Union under the Soviet-Polish Treaty on State Frontier (Moscow, Aug. 16, 1945), *UNTS*, 10:193.

67. *Sekai shuho*, Aug. 11, 1964 (Tokyo); reprinted in part in Doolin, *Territorial Claims in the Sino-Soviet Conflict*, (Stanford, 1965), p. 42.

ening peace and the security of the people." It further charged that Mao's statement would benefit only the imperialists, who are constantly attempting to sow mistrust and engender animosity among the peoples of socialist countries.[68]

(3) When the (Warsaw) Treaty of Friendship, Cooperation and Mutual Assistance was signed on May 14, 1955,[69] Communist China's observer to the drafting conference, General P'eng Teh-huai, said that he was authorized to declare his government's "full support and cooperation" with the treaty.[70] From then until 1961, Communist China sent an observer to every session of the Political Consultative Committee established under the treaty.[71] But its attitude toward the Warsaw Treaty changed with the deterioration of Sino-Soviet relations in the 1960's. A *People's Daily* editorial of September 20, 1968, supporting Albanian withdrawal from the Warsaw Treaty, stated:

> The Warsaw Treaty has long [been] an instrument for aggression in the hands of the Soviet revisionist renegade clique in pushing ahead with its social-imperialism. By using this treaty, this clique has put the national defense forces of the other member countries into its hands, arbitrarily stationing its own armed forces and holding military exercises in these countries, and it controls, enslaves and plunders them. The Warsaw Treaty controlled by the Soviet revisionist renegade clique is no different from NATO controlled by U.S. imperialism.[72]

These comments on the Warsaw Treaty may indicate that in the Commu-

68. See n. 65 above.
69. *UNTS*, 219:3. Parties to the treaty are Albania, Bulgaria, Hungary, East Germany, Poland, Romania, Czechoslovakia, and the Soviet Union.
70. *WCC*, 3:292.
71. See Grzybowski, *Socialist Commonwealth of Nations* (New Haven, 1964), p. 212. Since 1961, neither the authoritative *People's Daily* nor the *WCC* have contained information on Communist China's participation in activities conducted under the Warsaw Treaty.
72. "Courageous and Resolute Revolutionary Action," p. 1; English trans., same title, in *PR*, 11:9, no. 39 (Sept. 27, 1968). Albania withdrew from the Warsaw Treaty in accordance with a law passed by the Albanian People's Assembly 6th session, Sept. 12 and 13, 1968; *PR*, 11:7, no. 38 (Sept. 20, 1968). In a speech to the Assembly concerning Albanian withdrawal, Shehu, chairman of the Council of Ministers of Albania, said that the Soviet Union and other member states of the Warsaw Treaty had placed Albania "in a position of inequality and discrimination." He further charged that the Warsaw Treaty had "turned from a treaty of defense of the socialist countries against imperialist and revanchist aggression into a tool in the hands of the Soviet revisionists to liquidate the freedom of the peoples of the countries participating in this treaty," in "Warsaw Treaty Has Become Instrument for Soviet Revisionists' Aggression Against and Enslavement of the People of Member State," *PR*, 11:10, 13, no. 38 (Sept. 20, 1968).

nist Chinese view a treaty once equal or just can become unequal or enslaving as circumstances change.

(4) Though Communist Chinese leaders and scholars consider all treaties imposed upon China in the nineteenth and twentieth centuries unequal treaties, they have, until recently, refrained from specifically designating the 1858 Aigun Treaty and the 1860 Peking Treaty by which Czarist Russia acquired a large portion of Chinese territory adjacent to the present Chinese Manchuria.[73] With the intensification of the Sino-Soviet dispute, however, the Communist Chinese government has specifically pointed out that the two were unequal treaties.[74] It has cited the writings of Engels, Marx, and Lenin to support its position.[75] Such an assertion has of course been sharply denounced by the Soviet Union, which considers the Sino-Soviet border "the result of historical development over a long period," rather than the product of unequal treaties.[76]

73. *E.g.*, nowhere in the *Selected Works of Mao Tse-tung* (Peking, 1961–1965) is there mention of the Sino-Russian treaties of 1858 (Aigun) and 1860 (Peking).

74. See the *People's Daily* editorial mentioned above, n. 38. See also "Letter of the Central Committee of the C.P.C. of February 29, 1964, to the Central Committee of the C.P.S.U.," *PR*, 7:13, no. 19 (May 8, 1964).

75. *E.g.*, Communist China's statement of May 24, 1969: "The great teachers of the world proletariat Marx, Engels and Lenin had long made brilliant conclusions on the unequal nature of these treaties. Commenting on the 'Sino-Russian Treaty of Aigun' in 1858, Marx said that '. . . by his second opium war he [John Bull] has helped her [Russia] to the invaluable tract lying between the Gulf of Tartary and Lake Baikal, a region so much coveted by Russia that from Czar Alexey Michaelowitch down to Nicolaus, she had always attempted to get it.' Engels also pointed out in the same year that Russia despoiled 'China of a country as large as France and Germany put together, and of a river as large as the Danube' . . . Lenin pointed out that '. . . the European governments (the Russian Government among the very first) have already started to partition China. However, they have not begun this partitioning openly, but stealthily, like thieves,' and that 'The policy of the tzarist government in China is a criminal policy,'" in "Statement of the Government of the People's Republic of China, May 24, 1969," *PR*, 12:5, no. 22 (May 30, 1969).

76. *E.g.*, see Soviet statement of June 13, 1969; English trans. in *CDSP*, 21:9–13, no. 24 (July 9, 1969). One Soviet scholar ingeniously argued to support the Soviet position on the nature of the 1858 and 1860 Sino-Russian treaties concerning the Manchurian border, writing that the 1689 Sino-Russian Treaty of Nerchinsk (providing that the vast territory north of the Amur River belonged to China) was unequal, so the 1858 and 1860 Sino-Russian treaties merely rectified unjust situations and were therefore not unequal; Khvostov, "The Chinese 'Account' and Historical Truth," *Mezhdunarodnaya zhizn*, no. 10 (October 1964), English excerpt in Gittings, *Survey of the Sino-Soviet Dispute* (London, 1968), pp. 164–165.

V Interpretation of Treaties

Communist Chinese writer Wang Yao-t'ien has written that "treaty interpretation is a condition for clarifying the purpose, contents and correct application of a treaty . . . [and] is also a concrete condition for clarifying how a particular provision or the whole treaty can be applied to international relations."[1] Except for such brief comments as these and some general comments on certain Western practices that allegedly violate particular treaties, Communist Chinese writers rarely discuss the problem of treaty interpretation.

Many treaties concluded by Communist China, however, contain provisions concerning problems of interpretation, such as how to decide which text prevails in cases of discrepancies between Chinese and non-Chinese versions of the treaty, and what methods of settling treaty disputes shall be used. And the subject is obviously an important one.

In 1958, when many Communist Chinese writers engaged in controversy over the problem of systems of international law, Lin Hsin argued that there are two separate systems of international law – one bourgeois and the other socialist – which exist side by side.[2] He viewed international law as a reflection of the common ideological values shared by the states subject to its rules; since no common values existed between socialist and bourgeois states, accordingly there could be no uniform rules of international law equally binding upon both. Lin Hsin admitted that relations do exist between socialist and bourgeois states, but he nevertheless refused to concede the existence of a common system of international law adjusting such relations. He stated that all treaties and agreements concluded between bourgeois and socialist states are reached only after a fierce struggle between the two parties resulting in a compromise which reflects not common values but the realities of the power balance between them. The degree to which agreements are possible depends upon the extent to which the two systems of international law are capable of incorporating the new rule. The situation is one of "sleeping in the same bed but dreaming different dreams." When it comes to implementing a document of international law such as a treaty, agreement, or declaration, the struggle between the viewpoint of bourgeois international law

1. Wang Yao-t'ien, *Kuo-chi mao-yi t'iao-yüeh ho hsieh-ting* (International trade treaties and agreements) (Peking, 1958), p. 14.

2. "On the System of International Law after the Second World War," *CHYYC*, 1:34–38 (1958). For a study of the question of systems of international law, see summary of Communist writers' views in Hungdah Chiu, "Communist China's Attitude toward International Law," *AJIL*, 60:251–257 (1966).

and the viewpoint of socialist international law still continues. Such a struggle, in his view, is clearly reflected in different or even totally contrary interpretations of the same document. In other words, Lin seems to suggest that two different interpretations of a treaty are possible, depending upon the viewpoint of the interpreter.

Lin's view drew severe criticism from another writer, Chou Fu-lun. Chou gave an affirmative answer to the question whether there is a single system of international law exerting binding force upon all countries alike. He arrived at this conclusion by first stating that during a period of peace political, economic, and cultural relationships inevitably arise among all states, whatever their social system. Whereas Lin in holding that all law, including international law, reflects ideological values was forced to propound a two-system theory of international law, Chou set forth a different theory as to the nature and formation of international law: "International law should not be confused with municipal law. The major unique trait of international law is that its standards are formulated not by a super-legislature, but through agreement reached by the process of struggle, cooperation, compromise, and consultation." Chou felt that this common system of international law is constantly undergoing change as a result of the growth of new centers of power whose policies must be accommodated within the framework of international law. Consequently, current international law to him is neither bourgeois nor socialist, but instead reflects the transition from capitalism to socialism.

Chou was especially critical of Lin's idea that "when interpreting a common document of international law, the interpretations given by socialist and capitalist states are fundamentally different." He wrote:

> When imperialist countries engage in aggression, they often make distorted interpretations of international law to defend their aggressive acts. There are numerous cases in which they have described aggression as "self-defense," or armed intervention as a "police action" . . . If one adopts comrade Lin Hsin's position of the concurrent existence of two mutually opposed systems of international law, then in the case of executing a common document of international law, it is possible to make different interpretations of the document according to the mutually opposed viewpoints. This is to say that they have their reasons of international law and we have our reasons of international law. Finally, we can only reach the ambiguous conclusion of "you say you are right, and I say I am right."[3]

3. "On the Nature of Modern International Law–A Discussion with Comrade Lin Hsin," *CHYYC*, 3:52–56 (1958).

For Chou, then, there is only one correct interpretation of a treaty concluded between a socialist state and a bourgeois state. His position seems to coincide with the official Communist Chinese viewpoint.

Numerous Communist Chinese diplomatic documents have cited treaty violations by Western states regardless of the fact that these "violations" might be justified by different interpretations of the treaties.[4] Moreover, other Communist Chinese writers have adopted the same attitude. Wei Liang, for example, after analyzing many post-World War II treaties, reached the following conclusion:

> Imperialists basically do not pay attention to the treaty obligations they have assumed. When negotiating and concluding treaties with socialist countries, they sometimes have to agree to incorporate certain provisions which are consistent with norms of modern international law. However, they always try to distort these provisions and sometimes even attempt to undermine them publicly. The former President of the United States, Wilson, once publicly said that treaties cannot be interpreted literally but should be interpreted in accordance with the prestige, position, and will of the contracting parties. This is a frank confession of the arbitrary attitude of imperialism that treaties can be interpreted at will or, in other words, violated at will.[5]

The language or languages used in a treaty are the choice of the contracting parties. Nevertheless, Communist Chinese writer Wang Yao-t'ien has written that, in accordance with the principle of equality of the languages of the contracting states, bilateral treaties should use the languages of both contracting states and both texts should be equally authentic.[6] In Communist Chinese treaty practice, a treaty usually has two authentic texts, one in Chinese and one in the language of the other con-

4. *E.g.*, see Communist Chinese statement demanding withdrawal of United States forces from Lebanon, July 16, 1958, *WCC*, 5:132; English trans. in "The Chinese Government Demands Withdrawal of U.S. Forces from Lebanon," *PR*, 1:7, no. 22 (July 22, 1958). Communist China charged that landing American forces in Lebanon "flagrant[ly] violat[ed] the fundamental principles of the United Nations Charter" despite the fact that the U.S. justified its action, inter alia, under art. 51 of the Charter relating to "collective self-defense"; see President Eisenhower's statement announcing the dispatch of United States forces to Lebanon, July 14, 1958, *DIA* (1958), p. 287. See also Potter, "Legal Aspects of the Beirut Landing," *AJIL*, 52:729 (1958); I Hsin, "What Does Bourgeois International Law Explain about the Question of Intervention?" *KCWTYC*, 4:48–49 (1960).

5. "On the Post Second World War International Treaties," in *Kuo-chi t'iao-yüeh chi* (International treaty series), 1953–1955 (Peking, 1961), p. 670.

6. Wang Yao-t'ien, *International Trade Treaties*, p. 14.

tracting party.[7] Some treaties include a third, generally in Russian or English.[8]

Of course, the question may arise as to how a treaty should be interpreted if there are differences among the several authentic texts. Some Communist Chinese treaties are silent on this question.[9] Others provide that the third text – English, French, or Russian – shall prevail.[10] Occasionally, a treaty with only two authentic texts provides that in case of a discrepancy between the two the non-Chinese text shall prevail.[11] The reason for making such an arrangement may be that the other contracting party does not have competent Chinese translators to verify the authentic Chinese version of the treaty.

Apparently Communist China considers "negotiation" the most appropriate method of settling questions of treaty interpretation. For example,

7. The Sino-Indonesian Treaty of Friendship (Apr. 1, 1961) provides in art. 8: "The present Treaty is done in duplicate in the Chinese and Indonesian languages, both texts being equally authentic," *TYC*, 10:7; English trans. in "Treaty of Friendship between the People's Republic of China and the Republic of Indonesia," *PR*, 4:11, no. 24 (June 16, 1961). In some treaties the other contracting party does not wish to use its own language as an authentic text; *e.g.*, the Sino-Burmese Treaty of Friendship and Mutual Non-Aggression (Jan. 28, 1960) does not have an authentic Burmese text. The treaty provides that it was done "in duplicate . . . in the Chinese and English languages, both texts being equally authentic," *TYC*, 9:44; English trans. in "Treaty of Friendship and Mutual Non-Aggression between the People's Republic of China and the Union of Burma," *PR*, 3:13, no. 5 (Feb. 2, 1960). Communist China, on the other hand, has always insisted that an authentic text be drawn up in the Chinese language. The Nationalist government in most cases has also insisted upon use of the Chinese language for one of the authentic texts of any of its treaties. Chinese treaties concluded prior to the ascent of the Nationalists in 1928 did not always provide for this.

8. *E.g.*, the Sino-Nepalese Treaty of Peace and Friendship (Apr. 28, 1960) was "done in duplicate . . . in the Chinese, Nepali, and English languages, all texts being equally authentic," *TYC*, 10:113.

9. *E.g.*, the Sino-Ghanaian Treaty of Friendship (Aug. 18, 1961) does not mention the problem of discrepancies between the Chinese and English texts; *TYC*, 10: 17.

10. *E.g.*, the Sino-Albanian Agreement on Cooperation in Technology and Technical Science (Oct. 14, 1954) was done "in Chinese, Albanian, and Russian, all texts being equally authentic"; but the Russian text was said to prevail in case of discrepancy in interpretation; *TYC*, 3:173. See also the Sino-Finnish Trade Agreement (June 5, 1953), *TYC*, 2:35, 37; and the Sino-Syrian Trade Agreement (Nov. 30, 1955), *TYC*, 4:118, 120. For a study of the languages of Communist Chinese trade agreements see Hsiao, "Communist China's Trade Treaties and Agreements," *Vanderbilt L.R.*, 21:630–631 (October 1968).

11. *E.g.*, the Sino-Yemen Arab Republic Friendship Treaty (July 6, 1964) provides that the Chinese and Arabic texts are equally authentic, but in case of a discrepancy in interpretation the Arabic text "shall prevail," *TYC*, 13:5.

Article 4 of the Sino-Nepalese Treaty of Peace and Friendship,[12] signed on April 28, 1960, provides: "Any difference or dispute arising out of the interpretation or application of the present treaty shall be settled by negotiation through normal diplomatic channels." The rationale for preferring "negotiation" as the most appropriate method for treaty interpretation is explained by Wang Yao-t'ien: "Since the subjects of international treaties are sovereign states, there cannot be a supra-national organ in international affairs to interpret international treaties and compel the contracting parties to accept its interpretation. Consequently, the interpreters of international treaties can only be the contracting states themselves, and the best method of settling this problem is through diplomatic negotiation."[13] This attitude conforms to the Soviet theory that "direct diplomatic negotiations should be viewed as the 'most correct manner' in the interpretation of treaties; it is, by the way, the most common in international practice."[14]

Some friendship treaties do not contain any provision expressly relating to interpretation, but they do embody a provision calling for pacific settlement of "any dispute." For instance, Article 2 of the Sino-Afghan Treaty of Friendship and Non-Aggression, signed on August 26, 1960, provides, inter alia, that the two parties "undertake to settle all disputes between them by means of peaceful negotiation without resorting to force."[15] It seems reasonable to assume that this broadly-worded proviso was meant to include disagreements concerning the interpretation of the treaty itself, as well as other types of disputes between the two countries.

12. *TYC*, 10:13. Similar provisions have been incorporated in a number of treaties. *E.g.*, art. 5 of Sino-Burmese Treaty of Friendship and Mutual Non-Aggression (Jan. 28, 1960), *TYC*, 9:44; art. 6 of Sino-Cambodian Treaty of Friendship and Mutual Non-Aggression (Dec. 19, 1960), *TYC*, 9:25; art. 13 of Sino-Indonesian Treaty on Dual Nationality (Apr. 22, 1955), *TYC*, 8:12; art. 20 of Sino-Indonesian Agreement on Air Communication (Nov. 6, 1964), *TYC*, 13:368; art. 8 of Sino-Congolese (B) Agreement on Maritime Transport (Oct. 2, 1964), *TYC*, 13:366. Provisions concerning interpretation of treaties or agreements do not appear in any agreements concluded between Communist China and other Communist countries.

13. Wang Yao-t'ien, *International Trade Treaties*, p. 14.

14. Triska and Slusser, *The Theory, Law, and Policy of Soviet Treaties* (Stanford, 1962), p. 112.

15. *TYC*, 9:12. Art. 5 of Sino-Indonesian Treaty of Friendship (Apr. 1, 1961) provides: "Should any dispute arise between the Contracting Parties, they shall settle it by consultation through diplomatic channels and other ways agreed upon by both Parties in a spirit of fraternal and sincere friendship," *TYC*, 10:7, English trans. in "Treaty of Friendship between the People's Republic of China and the Republic of Indonesia," *PR*, 4:11, no. 24 (June 16, 1961). Art. 2 of Sino-Ghanian Treaty of Friendship (Apr. 18, 1961) contains this statement: "The Contracting

Under the terms of the Korean Armistice Agreement of July 27, 1953,[16] the Commander of the Chinese People's Volunteers agreed to set up two commissions to supervise the performance of the agreement and to oversee the repatriation of prisoners of war. The first commission – the Neutral Nations Supervisory Commission, composed of two Communist countries (Czechoslovakia and Poland) and two neutral countries (Switzerland and Sweden) – was to exercise its functions by the rule of unanimity.[17] The Neutral Nations Repatriation Commission, comprised of the same four countries plus India, was to exercise its functions by majority rule.[18] A few months after the signing of the Armistice Agreement, in a discussion of the means appropriate to supervising a cease-fire in Indo-China, Premier Chou En-lai commented on the two commissions:

> In the discussion there remains another question, namely, whether the neutral nations supervisory commission [to be created to supervise the Indo-China cease-fire] should adopt the principle of unanimity. Some people are of the opinion that the method of majority vote in the neutral nations supervisory commission would be adequate to settle questions. They are against the adoption of the principle of unanimity. The Delegation of the People's Republic of China cannot agree [with] this point. We hold that in present-day international affairs, the principle of unanimity is [the] most impartial and most reasonable principle which is best capable of settling important questions, whereas the method of majority vote has often been used on important international questions as an instrument for attempting to impose the will of the majority side of states on the minority side of states . . . Some people say that the Neutral Nations Supervisory Commission in Korea has been paralysed because it follows the principle of unanimity. That is an erroneous assertion. The fact is that the Neutral Nations Supervisory Commission in Korea has been effective in carrying out its main functions in accordance with the Armistice Agreement. In the 10 months after the Korean armistice, the Neutral Nations Supervisory Commission has supervised and examined the entry into and exit from

Parties will settle all disputes between them by means of peaceful negotiation," *TYC*, 10:17, English trans. in "Treaty of Friendship," *PR*, 4:7, no. 34 (Aug. 25, 1961). Similar provisions appear in art. 2 of Sino-Yemen Arab Republic Treaty of Friendship (June 9, 1964), *TYC*, 13:5; art. 2 of Sino-Guinean Treaty of Friendship (Sept. 13, 1960), *TYC*, 10:1; art. 4 of Sino-Malian Treaty of Friendship (Nov. 3, 1964), *TYC*, 13:381; art. 4 of Sino-Tanzanian Treaty of Friendship (Feb. 20, 1965), *People's Daily*, Feb. 21, 1965, p. 1, English trans. in "Sino-Tanzanian Treaty of Friendship," *PR*, 8:9, no. 9 (Feb. 26, 1965).

16. *TYC*, 2:382; *UST*, 4:234; *TIAS* 2782.

17. See art. 2, C, sec. 46, of the Armistice Agreement.

18. See art. XI, sec. 24, of Annex on Terms of Reference on Neutral Nations Repatriation Commission to art. 3, sec. 51b of the Armistice Agreement.

> Korea of over two million military personnel of the two sides to the armistice and more than 7,000 combat aircraft of the U.S. side and has thereby enabled the armistice situation in Korea to remain unaffected up to now. How can it be said that the Neutral Nations Supervisory Commission in Korea is not effective? . . . There is still another kind of example. The Neutral Nations Repatriation Commission in Korea operated with the method of majority vote. But what was the result? I have stated twice that the important decision on the disposition of prisoners of war agreed upon by the Indian, Polish and Czechoslovak members, was not respected by the members who were in the minority, and was not carried out by the United Nations Command side. As a result, a deadlock was created in which the U.S. side forcibly retained more than 21,000 Korean and Chinese captured personnel, a deadlock unresolved up to now. . . It is clear that the experience of the Korean armistice does not bear out the assertion that the principle of unanimity would inveitably lead to deadlocks while the method of majority vote would definitely not. As to deadlocks, no matter whether the principal of unanimity or majority vote prevailed, they have all been caused by the violation on the part of the U.S. side of certain terms of the Armistice Agreement in Korea.[19]

The armistice agreements concerning Vietnam, Cambodia, and Laos, later signed under the auspices of the Geneva Conference, all provided for the establishment of an International Commission for Supervision and Control composed of representatives from Canada, India (chairman), and Poland.[20] Normally, this commission would operate by majority vote, but the unanimity rule would apply in the case of actual or threatened armistice violations which might lead to a resumption of hostilities. Failure of the commission to agree, or refusal of one of the parties to put its recommendations into effect, would be a matter for referral to the members of the Geneva conference.

19. Premier Chou's statement of June 9, 1954, concerning the Indochina question at the Geneva Conference, *WCC*, 3:71–73; English trans. in "Statement by Chou En-lai on the Indo-China Question, June 9, 1954," *PC*, 13:8–9, supp. (July 1, 1954). During the Korean War, U.N. forces captured about 20,000 Communist Chinese prisoners of war; when the Armistice Agreement was signed in 1953, three-fourths of them refused to return to Communist China. In January 1954 the U.N. Command released them, and they chose to go to the Republic of China; see Calvocoressi, *Survey of International Affairs 1953* (London and New York, 1956), pp. 224–230.

20. See arts. 20, 21 of Agreement on the Cessation of Hostilities in Cambodia (July 20, 1954), *BFSP*, 161:595 (1954); arts. 34, 35 of Agreement on the Cessation of Hostilities in Laos (July 20, 1954), *ibid.*, p. 758; arts. 41, 42 of Agreement on the Cessation of Hostilities in Viet Nam (July 20, 1954), *ibid.*, p. 834.

Communist China was not a party to the armistice agreements because it was not a belligerent; but it signed the 1954 Final Declaration of the Geneva Conference,[21] which provided, inter alia, that the members of the conference agree to consult each other on any question referred to them by the International Commission in order to insure respect for the agreements on the cessation of hostilities.

In 1962 Communist China signed both the Declaration[22] and the Protocol on the Neutrality of Laos.[23] Article 1 of the Protocol said that the International Commission on Laos, set up by the 1954 Geneva Armistice Agreement on Laos, was to be entrusted with the function of supervising the neutrality of Laos. Article 14 of the Protocol provided for operation of the Commission as follows:

> Decisions of the Commission on questions relating to violations of Articles 2, 3, 4, and 6 [withdrawal of foreign forces and the introduction of war materials into Laos] of the Protocol or of the cease-fire referred to in Article 9, conclusions on major questions sent to the Co-Chairmen and all recommendations by the Commission shall be adopted unanimously. On other questions, including procedural questions, and also questions relating to the initiation and carrying out [of] investigations (Article 15), decisions of the Commission shall be adopted by majority vote.

Contracts between Communist China and other Communist countries concluded under a trade or payment agreement are usually governed by a document called "general conditions for the delivery of goods," which usually provides for arbitration of all trade disputes if negotiation has failed to produce a solution. For example, Article 27 of the 1953 General Conditions for the Delivery of Goods between the [Communist] Chinese and [East] German Foreign Trade Organs states:

> All disputes arising from or connected with a contract, if the parties cannot reach agreement, shall be arbitrated before the following arbitration organs. Both parties must comply with and execute the award of the arbitral organs. (1) If the defendant is the foreign trade organ of China, the arbitration shall be held in an arbitration organ in Peking. (2) If the defendant is the foreign trade organ of Germany, the arbitration shall be held in an arbitration organ in Berlin. Both the

21. *TYC*, 3:14; *BFSP*, 161:359 (1954).
22. *TYC*, 11:3; *UST*, 14:1104; *TIAS* 5410; *UNTS*, 456:301.
23. *TYC*, 11:7; *UST*, 14:1129; *UNTS*, 456:324.

> vendor and the vendee shall respectively appoint two persons familiar with economic matters as arbiters and the fifth arbiter shall be selected by the arbiters appointed by both parties.[24]

According to information contained in the official Compilation of Treaties of the People's Republic of China, only a few trade agreements or semi-official trade agreements concluded between Communist China and non-Communist countries in the 1949–1964 period contained arbitration clauses.[25] The composition of the arbitration tribunal is not the same in all these agreements. In two semi-official agreements with Japan concluded on December 31, 1952, and October 29, 1953,[26] neither the composition of the tribunal nor the method of selecting the arbitrators is mentioned. However, in the Sino-Finnish Trade Agreement of June 5, 1953,[27] it is provided that each party appoint one arbitrator and the umpire be chosen by agreement of the two arbitrators so appointed. Both the arbitrators and the umpire must be citizens of either China or Finland. In the semi-official trade agreement concluded between the China Import and Export Corporation and the French Industrial and Commercial Trade Delegation on June 5, 1953,[28] it is provided that if the arbitration is conducted in France each party is entitled to appoint an arbitrator of French nationality, and an umpire is to be chosen by agreement of the arbitrators. If the arbitrators cannot agree, the umpire is to be chosen by the president of the Civil Court of Seine Province. On the other hand, if the arbitration is conducted in Peking, the prevailing rules are those of the Foreign Trade Arbitration Commission of the China Council for the Promotion of International Trade.[29]

In a semi-official trade agreement with Japan concluded on March 5, 1958, the following arbitration arrangement is stipulated:

> Article 8. All disputes arising either from the effectuation of, or in relation to, trade contracts shall be settled by negotiation between the

24. *TYC*, 2:149. Generally this kind of document only applies to trade with Communist countries, but Communist China has concluded a general conditions document with Finland which also contains arbitration clauses; see n. 27, below and accompanying text.

25. For a study of Communist China's means of settling trade disputes see Hsiao, "Communist China's Foreign Trade Contracts and Means of Settling Disputes," *Vanderbilt L.R.*, 22:503, no. 3 (April 1969).

26. *TYC*, 2:367, 369.

27. *Ibid.*, p. 35.

28. *Ibid.*, p. 377.

29. For a study of this organ see Hsiao, "Communist China's Foreign Trade Organization," *Vanderbilt L.R.*, 20:314–317, no. 2 (March 1967).

> individual contracting parties. When no settlement is agreed upon between them, such disputes shall be submitted for arbitration.
>
> The arbitration shall be rendered in the country where the defendant resides.
>
> In the case of arbitration in Japan, the International Trade Arbitration Association of Japan shall take charge of arbitration in conformity with the arbitrational regulations of the said Association. The nomination of the arbitrators shall not necessarily be limited to those on the arbitrators' list of the said Association. However, they shall be limited to the nationals of the third country agreed upon by both Contracting Parties, or by Japan and the People's Republic of China.
>
> When arbitration takes place in the People's Republic of China, such arbitration shall be conducted by the Foreign Trade Arbitration Committee within the China Committee for the Promotion of International Trade, in conformity with the arbitral regulations of the said Committee.
>
> The arbitrary decision shall be the final decision, and it shall be complied with by both Contracting Parties.
>
> Both Contracting Parties shall obtain consent from their Governments respectively, to accord all facilities for the fulfilment of the arbitration and the travelling of personnel required therefore, and for the guarantee of safety for such personnel.[30]

Another, similar form of arbitration clause is provided in the semi-official trade agreement between the China Council for the Promotion of International Trade and the Eastern Commission of German Economy of September 27, 1957.[31] The agreement designated Zurich, Switzerland, as the location of arbitration. Under a supplemental agreement concluded on the same day,[32] when one of the parties fails to appoint an arbitrator within a month, the local Chamber of Commerce can appoint an arbitrator for that party upon the request of the other party. Moreover, the local Chamber of Commerce can appoint the umpire upon the request of either party when the arbitrators themselves have failed to reach an agreement on appointment of the umpire. According to a study by Gene Hsiao, this form of arbitration clause is also "frequently used in Peking's foreign trade contracts with capitalist countries."[33]

Some trade agreements between non-Communist countries and Communist China do not contain arbitration clauses, but provide for the ap-

30. *TYC*, 7:197; English trans. in *Contemporary Japan* (Tokyo), 25:520 (1957–1959).
31. *TYC*, 6:323.
32. *Ibid.*, p. 328.
33. Hsiao, "Communist China's Foreign Trade Contracts," p. 518.

pointment of a mixed commission to supervise execution of the agreement. For instance, Article 12 of the 1964 Sino-Central African Republic Agreement on Exchange of Goods and Payment provides: "In order to expand trade relations between the two countries and to supervise the smooth execution of this agreement, the contracting parties shall establish a mixed commission to be composed of representatives appointed by the governments of both countries."[34] Presumably, this mixed commission also settles disputes under the agreement.[35]

In many treaties concluded between non-Communist countries it is provided that disputes on treaty interpretation are to be referred for decision by the International Court of Justice, or for arbitration, conciliation, or third-party settlement. Communist China's treaty practice does not conform to this practice. With the exceptions of trade agreements and the few cases mentioned above, Peking's treaties or agreements do not contain any such arrangement for treaty interpretation. This attitude is similar to the Soviet attitude, which opposes any reference of disputes on interpretation (with the exception of disputes on interpretation of trade agreements) to adjudication or third-party settlement.[36]

Communist Chinese writers do not seem to have discussed the question of rules governing treaty interpretation, a topic that is dealt with in most Western treatises on international law. Moreover, although Communist China and its writers frequently charge other states with violating treaties, their charges are usually vague and general and do not contain sophisticated legal arguments. It is therefore difficult to sort out from these charges the rules that Communist China considers applicable to treaty interpretation.

Nevertheless, in two cases Communist China has engaged in lengthy debate with its treaty partners on the interpretation of certain provisions. The first is the Sino-American Agreed Announcement on Repatriation of Civilians, signed on September 10, 1955.[37] In this Announcement, Communist China "recognizes that Americans in the People's Republic of

34. *TYC*, 13:262.

35. The 1964 Sino-Algerian Trade Agreement explicitly provides this; according to art. 14 of the Agreement: "At the request of either contracting party, the mixed commission shall be convened within two months of the notification of the request, to study and to promote trade exchange and to settle any possible difficulties and disputes which may arise out of the execution of this agreement," *TYC*, 13:287.

36. Triska and Slusser, *Soviet Treaties*, pp. 382–383, 388.

37. "U.S., Red China Announce Measures for Return of Civilians," *DSB*, 33:456, no. 847 (Sept. 19, 1955).

China who desire to return to the United States are entitled to do so." According to the United States' view, explained in a statement issued by the Department of State on December 15, 1955: "This declaration is simple, clear, and positive. It says that any United States citizen has the right to leave China . . . No distinction is made as between those in prison and those out of prison."[38]

The American interpretation of the Announcement was rejected by Communist China in a statement issued on January 6, 1956. It characterized the American position as a distortion of the Announcement, claiming that there was a distinction between "ordinary American residents" and "those who offended against law." The latter group would have to be dealt with in accordance with Chinese legal procedure; "only when they have completed their sentences or when China has adopted measures to release them before the completion of their sentences [could] the question of their exercising the right to return . . . arise."[39] Communist China also claimed that this distinction between two different groups of Americans in China had been recognized by the United States during negotiation of the Announcement at Geneva. It argued further that the question of what measures should be taken with respect to Americans in Chinese prison was within Chinese sovereignty and was not a matter for intervention by any other states.

A Communist Chinese article elaborating the negotiating history of the Announcement states:

> During the Geneva talks . . . [t]he U.S. . . . demanded the release of all American nationals, including spies and lawbreakers, within a specified time. China of course could not agree to this. A country has the inviolable sovereign right to deal with law-breaking aliens according to its own laws. American nationals violating Chinese law in China must be dealt with under Chinese law. *The American representative, Ambassador Johnson, finally had to agree to this principle* [emphasis added]. Agreement on this item was therefore reached on September 10, 1955.[40]

The United States does not seem to have officially challenged the Communist Chinese version of the negotiating history. But, according to

38. "Continued Detention of U.S. Civilians by Communist China," *DSB*, 33: 1049–1050, no. 861 (Dec. 26, 1955).

39. *WCC*, 4:2–3. See also Young, *Negotiating with the Chinese Communists: The United States Experience, 1953–1967*, (New York, 1968), p. 81.

40. Ho Yang, "U.S. Stonewalling at Geneva," *PC*, 16:13–14 (Aug. 16, 1956).

Kenneth Young's study, Ambassador Johnson "had never agreed to the 'principle,' alleged by Peking, that Americans in China were divided into two categories."[41] Communist China, however, has insisted upon its interpretation and has continued to hold several Americans in Chinese prison since the conclusion of the Announcement.

Communist China alleged that the United States government had itself violated the Announcement by failing to supply the name list of Chinese residents in the United States, and by failing to take any steps in regard to Chinese imprisoned in the United States comparable to those China took (supplying a name list and releasing some Americans from prison) with respect to Americans who had committed offenses in China.[42] The United States did not respond directly, but claimed that during negotiation of the Announcement Communist China had not mentioned Chinese convicts imprisoned in the United States.[43] Subsequent Communist Chinese statements concerning the Announcement did not mention whether, during negotiations, both sides had discussed the questions of supplying a name list or releasing Chinese convicts.

The merits of these disputed questions lie outside the scope of this book. But the exchange of communications concerning the Announcement reveals several interesting points concerning treaty interpretation. In the first place, it is a generally accepted rule of treaty interpretation that, when the text of a treaty is ambiguous or obscure, the contracting parties may resort to preparatory work (*travaux préparatoires*) to ascertain the meaning of the treaty.[44] During the Sino-American exchange of communications concerning the Announcement, Communist China frequently resorted to negotiating history to support its interpretation – a practice which surely suggests that it believes a contracting party may resort to preparatory work in treaty interpretation.

Another interesting point derives from Communist China's statement

41. Young, *Negotiating with the Chinese Communists*, p. 86. According to Young (p. 87), the State Department issued a statement on Jan. 29, 1957, which claimed, inter alia, that no distinction between Americans in or out of prisons in Communist China was ever made or implied by either side during negotiation of the "Announcement." But I have not been able to locate such a claim in the statement mentioned. See "Failure of Chinese Communists to Release Imprisoned Americans," *DSB*, 36:261–263, no. 921 (Feb. 18, 1957).

42. Statement of Communist Chinese Foreign Ministry, Mar. 31, 1956; *WCC*, 4:53.

43. See statement of the Department of State, Jan. 27, 1957; "Failure of Chinese Communists," *DSB*, 36:261–263, no. 921 (Feb. 18, 1957).

44. See Oppenheim, *International Law*, I, 8th ed. (London, 1955), 957. See also art. 32 of 1969 Vienna Convention on the Law of Treaties; *AJIL*, 63:885 (1969).

that the Announcement could not be interpreted to infringe upon its "sovereign right" to handle American law-breakers in accordance with Communist Chinese legal process. This seems to invoke the principle of *in dubio mitius* (in doubtful cases, the mild [one commands]) suggested by some Western writers on treaty interpretation. That is, Communist China's practice seems to agree with the rule that when "the meaning of the term is ambiguous, that meaning is to be preferred which is less onerous for the party assuming an obligation, or which interferes less with the territorial and personal supremacy of a party, or involves less general restrictions upon the parties."[45]

Finally, Communist China appears to insist upon the principle of reciprocity in treaty interpretation. Thus, when it provided a name list of Americans (in and out of Chinese prison) to the United States, it immediately demanded that the United States provide a name list of Chinese (in and out of American prison) in the United States, despite the fact that the preparatory work did not indicate that the United States had ever agreed to this arrangement. The Communist Chinese demand would, in fact, make treaty obligations between the United States and Communist China unequal: at the time of concluding the Announcement fewer than 100 Americans were in China,[46] while there were more than 200,000 Chinese in the United States. Moreover, most Chinese in the United States are pro-Nationalist and oppose the supplying of their names to Communist China, which they fear may subject their relatives in mainland China to Communist Chinese pressure. The Communist Chinese insistence on applying the principle of reciprocity to treaty interpretation does not seem to have any support in the writings of Western international law scholars.

In the second case – the Sino-Russian Additional Treaty of Peking, signed on November 14, 1860 – Article 1 of the Treaty describes the Russo-Chinese boundary in Manchuria as follows:

> Hereafter, the eastern boundary of the two Empires, begins at the confluence of the Shilka and Argun Rivers and shall follow the course of the Amur River to the point [where it] joins the Ussuri River. The land[s] located on the left (to the north) of the Amur River belong

45. Oppenheim, *International Law*, I, 953.

46. According to Communist China, at that time there were 59 Americans in China and 40 in Chinese prisons. See Statement of the Spokesman of Chinese Foreign Ministry, Jan. 18, 1956; *WCC*, 4:13.

> to the Russian Empire; and the lands located on the right (to the south) of the river, to the junction on the Ussuri River, belong to the Chinese Empire. Farther on, from the junction of the Ussuri River to the Hsingkai Lake, the boundary line runs along the Ussuri and Sungacha Rivers. The lands located east of these rivers (right) belong to the Russian Empire and the lands located west (left) of the rivers belong to the Chinese Empire.[47]

According to Communist China, since the treaty does not provide that the Amur and the Ussuri Rivers belong to Russia, the general principle of international law demarcating boundary rivers should govern the boundary line in these two rivers. That is to say, the main channel should form the boundary line of the two countries in these navigable rivers.[48]

The Soviet Union argues that in accordance with a map exchanged between China and Russia in 1861 and attached to the 1860 Treaty, the boundary line appearing in the map runs along the Chinese bank of the rivers in certain places, giving the Soviet Union title to the entire rivers in those places. To support its argument, the Soviet Union cites an 1858 treaty between Costa Rica and Nicaragua stating that the boundary line runs along the right (Costa Rican) bank of the San Juan River, giving Nicaragua exclusive possession of the river.[49]

The Soviet interpretation of the Treaty was categorically rejected by Communist China, as illustrated by the following Foreign Ministry statement:

> The attached map is on a scale smaller than 1:1,000,000. The red line on it only indicates that the rivers form the boundary; it does not, and cannot possibly show the precise location of the boundary line in the rivers . . . In order to deny the principle of international law that the central line of the main channel shall form the boundary line in the case of navigable boundary rivers, the Soviet Government cited as an example the treaty concluded between Costa Rica and Nicaragua in 1858 . . . it impudently alleged that the "Sino-Russian Treaty of Peking" was likewise a case in point. Of course, there are exceptions

47. *Treaties, Conventions* [etc.] *between China and Foreign States* (Shanghai, 1908), I, 36.

48. Information Department of the Chinese Foreign Ministry, "Chenpao Island Has Always Been Chinese Territory" (Mar. 10, 1969), *PR*, 12:14, no. 11 (Mar. 14, 1969).

49. Statement of the Soviet government, June 13, 1969; "Soviet Notes on Border Conflict With China," *CDSP*, 21:11, no. 24 (July 9, 1969). It is not clear whether the Soviet Union claims the whole of the Amur and Ussuri Rivers.

> to any established principle of international law, and the same is true of the principle that the central line of the main channel shall form the boundary in the case of navigable boundary rivers. *But explicit stipulations must be made in treaties for any exceptional case* [emphasis added] . . . Now we want to ask the Soviet Government: Where is it stipulated in the "Sino-Russian Treaty of Peking" that the boundary line between China and Russia runs along the Chinese bank of the [Amur] and [Ussuri] Rivers?[50]

This statement appears to invoke the principle *in dubio mitius* in interpretation of treaties mentioned before.[51] In other words: Communist China's position seems to be that, if the 1860 Sino-Russian Treaty explicitly provides that the Amur and Ussuri Rivers form the boundary line between China and Russia, it certainly cannot be interpreted to mean that the two rivers belong to Russia exclusively and thereby derogate Chinese sovereignty over these two rivers.

50. "Document of the Ministry of Foreign Affairs of the People's Republic of China--Refutation of the Soviet Government's Statement of June 13, 1969, October 8, 1969," *PR*, 12:14, no. 41 (Oct. 10, 1969).

51. See n. 44 above, and accompanying text.

VI Suspension and Termination of Treaties

A treaty may be terminated or suspended in several ways. Generally, no problem is created if this is done in accordance with the terms of the treaty or by consent of the parties to it. However, disputes can arise from termination or suspension on grounds of revolutionary change of government, vital change of circumstance (*rebus sic stantibus*), violation by a contracting state, or outbreak of war between the parties.

The internal law of each state determines which organ within the state is entitled to terminate or suspend a treaty. In Communist China, the 1954 Constitution provides in Article 31, paragraph 12, that the Standing Committee of the National People's Congress shall decide upon the ratification and abrogation of treaties. Similar provision is contained in Article 18 of the revised draft of the Constitution of the People's Republic of China adopted on September 6, 1970, by the Second Session of the Ninth Central Committee of the Chinese Communist Party.

The actual process by which Communist China terminates or suspends treaties is not clear. On a number of occasions, the Foreign Ministry reportedly notified the other contracting parties of the termination or suspension of certain treaties; but there was no indication whether this action was based on a decision of the Standing Committee of the National People's Congress or on a decision of some other organ such as the State Council.

If a state does not follow the internal process prescribed by its domestic laws for abrogating a treaty the question arises whether its action is internationally valid. For example, during the 1956 Hungarian incident, the question was posed with respect to the legality of Soviet intervention; the Soviets attempted to justify their action by reference to the Warsaw Treaty, which had already been denounced by the Nagy government. One Communist Chinese writer argued that the Hungarian Constitution allowed only the Presidium of the National Assembly to denounce treaties; since the Presidium had not acted, the Warsaw Treaty was still valid at the time of Soviet intervention.[1]

Treaties limited in time terminate when the specified period elapses unless they are renewed. Many Communist Chinese trade or payment agreements are effective for a fixed period only. For instance, Article 9

1. Ch'en T'i-ch'iang, "The Hungarian Incident and the Principle of Non-Intervention," *Enlightenment Daily*, Apr. 5, 1957, p. 1. The author was later declared a rightist and his article severely criticized by other Communist Chinese writers, but none of the criticism was directed against parts of the article dealing with this point.

of the Sino-Bulgarian Agreement on Exchange of Goods and Payment in 1953 provides: "The effective period of this Agreement begins on January 1, 1953 and terminates on December 31, 1953."[2]

Many agreements concluded by Communist China, in addition to specifically providing their duration, also contain provisions on automatic renewal and termination by notice. For instance, Article 6 of the Sino-Burmese Treaty of Friendship and Non-Aggression provides: "(2) The present Treaty will come into force immediately on the exchange of the instruments of ratification and will remain in force for a period of ten years. (3) Unless either of the Contracting Parties gives to the other notice in writing to terminate it at least one year before the expiration of this period, it will remain in force without any specified time limit, subject to the right of either of the Contracting Parties to terminate it by giving to the other in writing a year's notice of its intention to do so."[3]

Similar provisions have been included in several multilateral treaties participated in by Communist China. For example, Article 12 of the Agreement between [Communist] China, [North] Korea and the Soviet Union on Cooperation in Saving Lives and Aiding Ships and Aircraft in Distress at Sea, signed on July 3, 1956, provides: "The effective period of this Agreement is three years starting from January 1, 1957. If none of the contracting parties declares its abrogation of this Agreement six months before the expiration of the effective period, it shall remain effective for another year. Thereafter, if none of the parties declares its abrogation of this Agreement six months before the expiration of the effective period, each time it shall remain effective for another year."[4] On June 24, 1967, Communist China's Foreign Ministry notified the Soviet Embassy in Peking that it considered it unnecessary to extend this agreement; the agreement therefore presumably expired on January 1, 1968. On July 12, 1967, the Soviet Foreign Ministry sent a note to the Communist Chinese Embassy in Moscow expressing its "regret" at the Communist Chinese decision.[5] In the July 19, 1967, issue of *Izvestia* a Soviet professor of international law denounced the Communist Chinese

2. *TYC*, 2:95.
3. *TYC*, 9:44; English trans. in "Treaty of Friendship and Mutual Non-Aggression between the People's Republic of China and the Union of Burma," *PR*, 3:13, no. 5 (Feb. 2, 1960).
4. *TYC*, 5:199.
5. See *Pravda*, July 18, 1967, p. 2; *Izvestia*, July 19, 1967, p. 2, English trans. in "Peking Refuses to Cooperate," *CDSP*, 19:9, no. 29 (Aug. 9, 1967).

decision as opposed to the spirit of international cooperation.[6] The *People's Daily* replied: "Since the conclusion of the agreement, the Chinese side has all along fulfilled its obligations under it. Now that the agreement has expired, the Chinese government has every right to declare it as invalid, as is provided for by the provisions of the agreement. It is normal and unimpeachable to do so in international relations. By using this as a pretext to vilify China, the Soviet revisionist ruling clique [is] making trouble without the slightest justification."[7]

Reportedly, Communist China has terminated its participation in several other multilateral treaties, such as the 1956 Agreement on the Establishment of a Joint Institute of Nuclear Research[8] and the 1956 Agreement on Cooperation concerning Fishing, Oceanographic, and Limnological Research in the Western Part of the Pacific Ocean.[9] Both treaties explicitly authorized termination by any party upon appropriate notice.

A treaty which does not mention the duration of its effective period, or which stipulates a time period which has not yet expired, may nevertheless be terminated by mutual consent of the contracting parties. Consent may be expressed or implied. Thus, when the parties to a treaty conclude a new treaty covering the same objects as a former treaty, the earlier one is terminated by implied consent.[10]

Communist China appears to accept these principles. As an illustration, in 1958 it concluded a friendship treaty with Yemen which was to be effective for ten years.[11] Before expiration of this treaty, the two countries concluded a second friendship treaty[12] which nullified the 1958 treaty upon enforcement of the 1964 treaty.[13] On other occasions, abrogation

6. G. Zadorozhny, "Peking Opposes Cooperation," p. 2; English trans. in *CDSP*, 19:9–10, no. 29 (Aug. 9, 1967).

7. Commentator, "The Soviet Revisionist Renegades Try to Pull the Wool over the Eyes of the Public," *People's Daily*, Aug. 5, 1967, p. 6; English trans. in "Moscow Tries to Pull the Wool over the Eyes of Public, Says *Jen-min Jih-pao*," *SCMP*, 3997:47 (Aug. 9, 1967).

8. *TYC*, 10:408; *UNTS*, 259:125. All Communist states participated in this Agreement. In 1966, UPI reported that Communist China had withdrawn from the Institute; UPI news release, London, July 13, 1966; *China News* (Taipei), July 14, 1966. See also Zadorozhny, "Peking Opposes Cooperation."

9. *TYC*, 5:169. Parties to the agreement before Communist China's withdrawal: Communist China, North Korea, North Vietnam, Outer Mongolia, the Soviet Union. Communist China's withdrawal was reported in Zadorozhny, "Peking Opposes Cooperation."

10. Oppenheim, *International Law*, I, 8th ed. (London, 1955), 937–938.

11. Art. 5 of the treaty; *TYC*, 7:3.

12. *TYC*, 13:5.

13. The treaty came into force June 6, 1964, when it was signed (art. 4). When the 1958 treaty was concluded, Yemen was a kingdom; a republican regime was established in 1962 which until 1969 controlled only about half the country.

by mutual consent has been sanctioned by a document separate from the treaty. The 1945 Sino-Soviet Treaty of Friendship and Alliance, for example, was to have remained in force for a period of thirty years.[14] In 1950, however, Communist China and the Soviet Union concluded a new treaty of friendship, alliance, and mutual assistance[15] that did not mention the validity of the old treaty; but an exchange of notes between the parties observed that: "In view of the signature of the . . . Treaty [of Friendship, Alliance and Mutual Assistance] . . . and in accordance with the said treaty . . . the two Contracting Parties agree to declare that the Chinese-Soviet Treaty of Friendship and Alliance . . . concluded on 14 August 1945 between China and the Soviet Union . . . ha[s] ceased to have effect."[16]

Western scholars generally accept the principle that "[c]hanges in the government or in the constitution of a state have . . . no effect upon the continued validity of its international obligations."[17] This principle was first challenged by the Russian Communists soon after they came to power in 1917. The Soviet government repudiated in principle all treaties imposed on other states by the Czarist government and also all loan agreements concluded by the Czarist and provisional bourgeois governments of Russia. Simultaneously, it reaffirmed many treaties and agreements which in its view did not infringe on its rights or the rights of the other contractors as sovereign states.[18] However, the treaties under which the Czarist government had seized territories from other states (for example, Turkey and China) were not abolished.

14. *UNTS*, 10:300. The treaty was concluded by the Nationalist government. Communist China did not denounce this treaty when it came into power in 1949.

15. *TYC*, 1:1; *UNTS*, 226:12.

16. *UNTS*, 226:16.

17. Oppenheim, *International Law*, I, 948–949. O'Connell: "Change of Government does not affect the personality of the State, and hence a successor government is required by international law to perform the obligations undertaken on behalf of the State by its predecessor. This is true even when the change is revolutionary," O'Connell, *International Law* (London, 1965), I, 456. The 1969 Convention on the Law of Treaties does not cover this problem. The U.N. International Law Commission is now considering this problem in connection with its work on state succession.

18. Whether the Soviet Union is a new state or the continuation of the old Russian state is controversial among Soviet jurists; see Avakov, *Pravopreemstvo sovetskogo gosudarstva* (Moscow, 1961). Yevgenyev, a Soviet doctoral candidate, expressed the view that the Soviet Union is a state of a new historical type; in that case, treaties of the former state continue to bind a new state only in cases when, "firstly, the new State makes an official statement to this effect and, secondly, if there is no explicit statement regarding non-recognition or annulment of the treaty (tacit affirmation)," Yevgenyev, "The Subject of International Law," in Kozhevnikov, ed., *International Law* (Moscow [1961]), p. 126.

When the Communist Chinese took over mainland China in 1949, they reacted much as the Soviet government had with respect to treaties concluded by the Nationalist or other earlier Chinese governments. Article 55 of the Common Program of the Chinese People's Political Consultative Conference, which operated for five years as a provisional constitution until promulgation of a formal constitution in 1954, declared: "The Central People's Government of the People's Republic of China must study the treaties and agreements concluded by the Kuomintang government with foreign governments and, depending on their contents, recognize, annul, revise or re-conclude them."[19]

Communist China seems to believe that the state of China continues to exist, despite establishment of a new Communist regime and the 1949 change of name from "Republic of China" to "People's Republic of China." Chou Keng-sheng wrote in the *People's Daily* that "the People's Republic of China, established after the victory of the Chinese people's revolution, is the continuation of pre-liberation China and, therefore, is unequivocally entitled to China's legitimate seat in the United Nations and to all other rights pertaining to it," in "China's Legitimate Rights in the United Nations Must be Restored," *People's Daily*, Dec. 5, 1961, p. 5. Shao Chin-fu made a similar claim:"In its class nature, the People's Republic of China is a new state entirely different from old China, but it is not a new member in the family of nations. In international law she is not a new subject, but continues to exist as a member in international relations in the place of old China," in "The Absurd Theory of 'Two Chinas' and Principles of International Law," *KCWTYC*, 2:9 (1959).

A Soviet textbook of international law deals with Communist China's case in a section entitled "Succession Following the Replacement of a State of One Historical Type by That of Another"; the Soviet case is also discussed; see Yevgenyev, "The Subject of International Law," in Kozhevnikov, ed., *International Law*, pp. 125–127.

19. *WCC*, 1:1; English trans. in *The Important Documents of the First Plenary Session of the Chinese People's Political Consultative Conference* (Peking, 1949), pp. 19, 20. Similar provisions were included in the Agreement on Internal Peace handed to the Nationalist government Apr. 15, 1949; art. 18 provides: "Both sides agree that all treaties and agreements concluded with foreign states during the rule of the National Government at Nanking and other diplomatic documents and archives, open or secret, shall be handed over by the National Government at Nanking to the Democratic Coalition Government and examined by the Democratic Coalition Government. All treaties or agreements which are detrimental to the Chinese people and their state, especially those which are in the nature of selling out the rights of the state, shall be either abrogated, or revised, or new treaties and agreements shall be concluded instead, as the case may be," Mao Tse-tung, "Order to the Army for the Country-wide Advance, April 21, 1949," in *Selected Works of Mao Tse-tung*, (Peking, 1965), IV, 394.

When the Nationalist government came to power in 1928 it was not happy about the unequal treaties concluded by former Chinese governments, but it did not attempt to unilaterally repudiate them until after its initial efforts had failed. The Nationalist declaration June 16, 1928, stated: "For eighty years China has been under the shackles of unequal treaties. These restrictions are a contravention of the international law principle of mutual respect and sovereignty and are not allowed by any sovereign state . . . Now that the unification of China is being consummated,

Though Article 55 refers only to Kuomintang (that is, 1928–1949) treaties, Communist Chinese practice indicates that it also applies to treaties concluded from 1912 to 1928 by various Peking governments and even to treaties of the pre-1911 Imperial government. This interpretation is also supported by a page-one editorial of March 8, 1963, in the authoritative *People's Daily*: "At the time the People's Republic of China was inaugurated, our government declared that it would examine the treaties concluded by *previous Chinese governments* with foreign governments, treaties that had been left over by history, and would recognize, abrogate, revise or renegotiate them according to their respective contents." (Emphasis added.)

The status of any particular pre-1949 treaty prior to a public declaration of the Communist Chinese government is unclear. Following is a summary of Communist Chinese practice with respect to treaties concluded by former Chinese governments.

Multilateral Treaties

Communist China apparently regards treaties relating to the establishment of international organizations as continuing in force regardless of revolutionary changes in government. This inference is drawn from the fact that shortly after the establishment of the Communist regime on October 1, 1949, Communist China demanded to be seated in the United Nations, the specialized agencies of the United Nations, and the Allied Control Commission on Japan; it also demanded the immediate expulsion of the Nationalist delegation from each of these bodies.[20] Communist

---

we think the time is ripe for taking further steps and beginning at once to negotiate – in accordance with diplomatic procedure – new treaties on a basis of complete equality and mutual respect for each other's sovereignty," in "Declaration of the Nationalist Government, June 16, 1928," *Chinese Social and Political Science Review*, 12:47, doc. sec. (1928).

20. In a cablegram to the President of the General Assembly, Nov. 18, 1949, the Foreign Minister of the People's Republic of China stated that his government repudiated the legal status of the delegation of the National government of China headed by T. F. Tsiang, holding that it could not represent China and had no right to speak on behalf of the Chinese people in the U.N.; U.N. Doc. A/1123 (1949). In another cablegram, Jan. 8, 1950, to the President of the General Assembly and the Secretary-General, the Foreign Minister demanded the expulsion of delegates of the Republic of China from the Security Council; U.N. Security Council, off. rec., 5th yr., 459th mtg., 1:2 (Jan. 10, 1950). In a third cablegram, Jan. 20, 1950, he informed the Security Council, Secretary-General, and U.N. membership that his government had appointed Chang Wen-t'ien as chairman of its delegation to attend meetings and participate in U.N. work including meetings and work of the Security

China's assertion of rights was necessarily predicated on a recognition of the continuing validity of the treaties establishing these organizations.

Communist China has also recognized the validity of several agreements concluded by the Nationalist government during the Second World War, including the 1942 United Nations Declaration, the 1943 Cairo Declaration, the 1945 Potsdam Proclamation, and the 1945 Japanese Instrument of Surrender. This recognition can be inferred from the invocation of these agreements by Communist China in connection with the questions of Taiwan and the Japanese peace treaty.

As for other multilateral treaties, Communist China has thus far recognized the following treaties signed or concluded by the Nationalist Chinese government:

The Geneva Protocol of 1925 Prohibiting the Use in War of Asphyxiating, Poisonous or Other Gases, and of Bacteriological Methods of Warfare.[21] The Nationalist government acceded to this Protocol on August 7, 1929. On July 13, 1952, Foreign Minister Chou En-lai issued a statement recognizing this agreement;[22] apparently Communist China did not demand the replacement of the Nationalist instrument of accession de-

Council. He asked when the Kuomintang representative would be expelled from the U.N. and the Security Council, and when the delegation of the People's Republic of China could begin to participate; *WCC*, 1:91; *Yearbook of the United Nations, 1950* (New York, 1951), p. 424. Again in a cablegram, Aug. 26, 1950, the Foreign Minister recalled the previous notes sent by his government to the Secretary-General and the General Assembly calling for expulsion of the Kuomintang representatives from all U.N. organs; continued toleration of those representatives was, he declared, a violation of the U.N. Charter and involved disregard of the rightful claims of the People's Republic of China. He requested that the necessary arrangements be made for the delegation of the People's Republic of China to attend the 5th General Assembly session; UN Doc. A/1364 (1950). Similar cablegrams were sent to other U.N. organs and specialized agencies and to other intergovernmental and nongovernmental international organizations; see *WCC*, 1:96 (ECOSOC – Feb. 2, 1950), 111 (ITU – Mar. 29, 1950), 113 (International Red Cross Association – Apr. 28, 1950), 114 (ECAFE – Apr. 28, 1950), 114 (UPU – May 5, 1950), 118 (FAO – May 12, 1950), 119 (UNESCO – May 12, 1950), 119 (WHO – May 12, 1950), 120 (WMO – May 12, 1950), 126 (ICAO – May 30, 1950), 127 (Trusteeship Council – May 30, 1950), 128 (ILO – June 5, 1950), 128 (International Law Commission – June 6, 1950), 129 (Allied Control Commission in Japan – June 19, 1950), 137 (ITU Administrative Council – Aug. 26, 1950), 138 (UNICEF – Aug. 26, 1950), 139 (IMF – Aug. 26, 1950), 140 (IBRD – Aug. 26, 1950); *WCC*, 2:1 (UPU – Jan. 9, 1951), 5 (ECAFE – Jan. 18, 1951), 72 (UPU – May 17, 1952), 91 (ITU – Sept. 23, 1952).

21. *LNTS*, 94:65; Chinese accession, *ibid.*, 71.

22. *TYC*, 6:320; *WCC*, 2:78; English trans. of Chou's statement in "On China's Recognition of the Protocol of June 17, 1925 Prohibiting Chemical and Bacteriological Warfare," *PC*, 15:33 (Aug. 1, 1952). Chinese text of the protocol in *TYC*, 6:319.

posited with the French government. This, then, is one of the two multilateral treaties to which both Chinese regimes are parties.

The Four Geneva Conventions of 1949.[23] These conventions were signed in the name of China in 1949 by the Nationalist government, which has not yet ratified them. The signature of the Nationalist government was recognized by Communist China on July 13, 1952, and documents of ratification were deposited with the Swiss government on December 28, 1956.[24]

The 1930 Convention on Load Line.[25] The Nationalist government acceded to this convention on August 19, 1935. In 1957 Communist China recognized the convention and accepted it.[26] Here again, Communist China evidently did not demand the replacement of the Nationalist instrument of accession deposited with the government of the United Kingdom. This is the other multilateral treaty to which both Chinese regimes are parties.[27]

The International Regulations for Preventing Collisions at Sea, June 10, 1948.[28] The Regulations were signed by the Nationalist government in 1948 and were accepted by Communist China in 1957.[29]

23. *TYC*, 5:203, *UNTS*, 75:31 (wounded and sick in armed forces in the field); *TYC*, 5:231, *UNTS*, 75:85 (wounded, sick, and shipwrecked members of armed forces at sea); *TYC*, 5:255, *UNTS*, 75:135 (prisoners of war); *TYC*, 5:333, *UNTS*, 75:287 (civilian persons in time of war).

24. Foreign Minister Chou En-lai's statement (July 13, 1952), *WCC*, 2:78; English trans., "On China's Recognition of the 1949 Geneva Conventions," *PC*, 15:33 (Aug. 1, 1952). For Chinese ratification, see *UNTS*, 260:438, 440, 442, 444 (effective June 28, 1957).

25. *TYC*, 6:282; *LNTS*, 135:301; Nationalist Chinese accession, *LNTS*, 160:417.

26. *TYC*, 6:294. Decision to accept was adopted by the Standing Committee of the National People's Congress, 82nd meeting, Oct. 23, 1957.

27. The *Treaty Series* published by the British Foreign Office 1957–1962 did not list Communist China's Acceptance of the Convention. See "Supplementary List of Ratifications, Accessions, Withdrawals [etc.]," *TS*, no. 73 (1957), Cmd. 386; *TS*, no. 61 (1958), Cmd. 642; *TS*, nos. 39, 63, 76, 83 (1959), Cmd. 727, 828, 866, 933; *TS*, nos. 21, 48, 74, 87 (1960), Cmd. 1008, 1114, 1186, 1276; *TS*, nos. 23, 67, 98, 119 (1961), Cmd. 1346, 1455, 1530, 1627; *TS*, nos. 36, 47, 78 (1962), Cmd. 1711, 1806, 1897. Since 1963 the list published by the U.S. Department of State on contracting parties to the 1930 Loadline Convention includes both the People's Republic of China and the Republic of China; *Treaties in Force 1963* (Washington, 1964), p. 284.

28. *TYC*, 6:294; *UNTS*, 191:20.

29. Decision of the Standing Committee of the National People's Congress, Dec. 23, 1957, *TYC*, 6:313. The date for deposit of acceptance is not clear; according to Nationalist Chinese information, Communist Chinese acceptance was deposited with the United Kingdom government Jan. 27, 1958. The depository function of the U.K. government was transferred to the Inter-Governmental Maritime Consultative Organization (IMCO) July 13, 1959. Since China has been represented in IMCO

The position of Communist China with regard to other pre-1949 multilateral treaties concluded by the Nationalist government is unclear. Presumably, its silence indicates that those treaties are not considered binding upon Communist China – at least not yet.

Bilateral Treaties

Apparently, Communist China recognizes the continued binding force of all boundary treaties concluded by Chinese governments prior to 1949. In 1957, Premier Chou En-lai said: "It was the opinion of our government that, on the question of boundary lines, demands made on the basis of formal treaties [concluded by former Chinese governments] should be respected according to general international practice."[30] This statement does not imply, however, that "unequal" boundary treaties will be continued in force forever. Chou made it clear that recognition of the continuing validity of these treaties "by no means excluded the seeking by two friendly countries of settlement fair and reasonable for both sides through peaceful negotiation between their governments." Accordingly, while Communist China did not contest the validity of

---

by the Nationalist government, in a table listing states accepting the Regulations published by IMCO in November 1959 Communist China was not listed; *Li-fa-yüan kung-pao* (Gazette of the Legislative Yüan, Taipei), 34:15, no. 6 (Dec. 1, 1964). On Nov. 21, 1966, Nationalist Chinese acceptance was deposited with IMCO. According to U.K. information, Communist Chinese acceptance was deposited Jan. 29, 1959; *TS*, no. 39 (1959), Cmd. 727. In the list published in 1964 and in 1965 by the U.S. Department of State on contracting parties to the Regulations, Communist China was listed as a contracting party with this statement: "the United States does not recognize the so-called People's Republic of China as a state, and, therefore, it regards its adherence to this Convention as being without legal effect and attaches no significance thereto," *Treaties in Force 1964*, p. 258; *Treaties in Force 1965*, p. 262. When the new Regulations for Preventing Collisions at Sea – approved by the International Conference on Safety of Life at Sea, London, May 17–June 17, 1960 – came into force for the U.S. on Sept. 1, 1965, the 1948 Regulations terminated for the U.S. on that date; *Treaties in Force 1965*, p. 262 n. 1. Communist China has not yet acceded to the new Regulations.

In a collision incident between a U.S. ship and a Chinese fishing vessel which resulted in sinking of the latter on May 15, 1961, the Ningpo Marine Fisheries Company (China) invoked the 1948 Regulations against the Lykes Brothers Steamship Company (U.S.). Lykes denied responsibility; *People's Daily*, July 16, 1962, p. 1; English trans. in "Chinese Protest to U.S. Company on Sinking of Fishing Vessel," *SCMP*, 2782:33 (July 20, 1962).

30. Report on the Question of the Boundary Line between China and Burma to the 4th Session of the 1st National People's Congress, July 9, 1957, *WCC*, 4:343; English trans. in Chinese People's Institute of Foreign Affairs, *A Victory for the Five Principles of Peaceful Coexistence* (Peking, 1960), p. 19.

earlier treaties relating to the Sino-Burmese boundary,[31] it nevertheless concluded a new boundary treaty with Burma on January 28, 1960, which provided that "the new boundary treaty shall replace all old treaties and notes exchanged between the two countries."[32]

Since the early 1960's, Communist China has repeatedly charged that the treaties defining the boundaries of Hong Kong, Macao, and the Sino-Soviet border are unequal treaties and therefore subject to renegotiation at an appropriate time.[33] This view has drawn sharp attacks from the Soviet Union; while conceding that the boundary treaties relating to Hong Kong and Macao are subject to renegotiation because a large majority of people living in those areas are Chinese, a *Pravda* editorial of September 2, 1964, insisted that the Manchurian border with the Soviet Union must be distinguished because it "developed historically and was fixed by life itself, and [past] treaties regarding the border cannot be disregarded."[34]

31. Hertslet, *Hertslet's China Treaties* (London, 1908), I, 99 (1894 Treaty), 113 (1897 Treaty); *LNTS*, 163:177 (1935 Treaty); *TS*, no. 80 (1941), Cmd. 7246 (1941 Treaty).

32. *TYC*, 9:68; English trans. in Chinese People's Institute, *A Victory for the Five Principles*, p. 33.

33. In an editorial, Mar. 8, 1963 ("A Comment on the Statement of the Communist Party of the U.S.A.," p. 1), the *People's Daily* listed these Chinese boundary treaties as "unequal treaties": Treaty of Nanking of 1842, Treaty of Aigun of 1858, Treaty of Tientsin of 1858, Treaty of Peking of 1860, Treaty of Ili of 1881, Protocol of Lisbon of 1887, Treaty of Shimonoseki of 1895, Convention for the Extension of Hong Kong of 1898.

See also statement of the Chinese delegation at World Youth Forum meeting, Moscow, Sept. 23, 1964: "Hong Kong and Macao are Chinese territory occupied by British and Portuguese imperialists in accordance with unequal treaties. The Chinese people will recover them at an appropriate time," *People's Daily*, Sept. 27, 1964, p. 5. In an interview with a Japanese Socialist delegation, July 10, 1964, Chairman Mao said: "About a hundred years ago, the area to the east of [Lake] Baikal became Russian territory, and since then, Vladivostok, Khabarovsk, Kamchatka, and other areas have been Soviet territory. We have not yet presented our account for this list," in "Chairman Mao Tse-tung Tells the Delegation of the Japanese Socialist Party that the Kuriles Must Be Returned to Japan," *Sekai Shūhō* (Tokyo), Aug. 11, 1964, cited in Doolin, *Territorial Claims in the Sino-Soviet Conflict* (Stanford, 1965), p. 44.

34. "On Mao Tse-tung's Talk with a Group of Japanese Socialists," p. 2. English trans. in *International Affairs*, 10:80 (Moscow, 1964). This editorial stated: "We have declared and continue to declare that People's China has every right to press for the liberation and reunification of Taiwan and Hong Kong, which are a part of the country and the majority of whose population are Chinese." As for the Sino-Soviet boundary, the editorial said: "Whereas the borders of Czarist Russia were determined by the policy of imperialist mediators, the borders of the Soviet Union were formed as a result of a voluntary statement of the will of the peoples on the basis of the principle of free-determination of nations."

Communist China has rejected this Soviet argument. A statement issued by its Foreign Ministry on March 10, 1969, reiterates the Chinese position, as expressed in the 1964 Sino-Soviet negotiation on boundary questions, as follows:

> In 1964 the Chinese Government held boundary negotiations with the Soviet Government, during which the Chinese side made it clear that the [1858] "Sino-Russian Treaty of Aigun," the [1860] "Sino-Russian Treaty of Peking" and other treaties relating to the present Sino-Soviet boundary were all unequal treaties Czarist Russian imperialism imposed on China when power was not in the hands of the people of China and Russia.
>
> But, prompted by the desire to strengthen the revolutionary friendship between the Chinese and Soviet peoples, the Chinese side was willing to take these treaties as the basis for determining the entire alignment of the boundary line between the two countries and for settling all existing questions relating to the boundary.[35]

The Soviet Union's statement of March 29, 1969, rejected the Chinese assertion that the several Sino-Soviet boundary treaties were unequal

---

Moreover, a letter of the Central Committee of the Communist Party of the Soviet Union to the Central Committee of the Communist Party of China explained: "Statements have recently been made in China concerning the aggressive policy of the Czarist government and the unjust treaties imposed upon China. Naturally, we will not defend the Russian Czars who permitted arbitrariness in laying down the state boundaries with neighboring countries . . . But . . . we cannot disregard the fact that historically formed boundaries between states now exist. Any attempt to ignore this can become the source of misunderstandings and conflicts; at the same time, they will not lead to the solution of the problems. It would be simply unreasonable to create territorial problems artificially at the present time, when the working class is in power and when our common aim is Communism, under which states' borders will gradually lose their former significance," in "Letter of the Central Committee of the C.P.S.U. of November 29, 1963, to the Central Committee of the C.P.C.," *PR*, 7:21, no. 19 (May 8, 1964). The Communist Party of China replied: "Although the old treaties relating to the Sino-Russian boundary are unequal treaties, the Chinese Government is nevertheless willing to respect them and take them as the basis for a reasonable settlement of the Sino-Soviet boundary question. Guided by proletarian internationalism and the principles governing relations between socialist countries, the Chinese Government will conduct friendly negotiations with the Soviet Government in the spirit of consultation on an equal footing and mutual understanding and mutual accommodation. If the Soviet side takes the same attitude as the Chinese Government, the settlement of the Sino-Soviet boundary question, we believe, ought not to be difficult, and the Sino-Soviet boundary will truly become one of lasting friendship," in "Letter of the Central Committee of the C.P.C. of February 29, 1964, to the Central Committee of the C.P.S.U.," *PR*, 7:13, no. 19 (May 8, 1964).

35. Information Department of the Chinese Foreign Ministry, "Chenpao Island Has Always Been Chinese Territory," *PR*, 12:15, no. 11 (Mar. 14, 1969).

treaties and invoked the 1924 Agreement on the General Principles for the Settlement of the Questions between the Republic of China and the Soviet Union[36] to support its view that all unequal treaties between China and Czarist Russia were abrogated.[37]

In a statement issued on May 24, 1969, Communist China also invoked the same Agreement to support its view that the existing Sino-Soviet boundary treaties should be renegotiated:

> [The Agreement] stipulates that at the conference agreed upon by both sides, they are "to annul all Conventions, Treaties, Agreements, Protocols, Contracts, etcetera, concluded between the Government of China and the Tzarist Government and to replace them with new treaties, agreements, etcetera, on the basis of equality, reciprocity and justice, as well as the spirit of the Declarations of the Soviet Government of the years of 1919 and 1920" and "to re-demarcate their national boundaries . . . and pending such redemarcation, to maintain the present boundaries."
>
> In pursuance of the 1924 Agreement, China and the Soviet Union held talks in 1926 to discuss the re-demarcation of the boundary and the conclusion of a new treaty. Owing to the historical conditions at the time, no agreement was reached by the two sides on the boundary question, no re-demarcation of the boundary between countries was made and no new equal treaty was concluded by the two countries.[38]

36. *LNTS*, 37:176.

37. See Soviet statement of Mar. 29, 1969, in *Pravda*, Mar. 30, 1969; English trans. in "Soviet Statement on Border Clashes Urges Negotiation," *CDSP*, 21:4–5, no. 13 (Apr. 16, 1969). According to this, the Soviet Union proposed to Communist China, May 17, 1963, that bilateral consultations on border questions be held; consultations began in February 1964 in Peking. The statement recalled that "the Soviet side presented proposals whose adoption permitted clarification through mutual agreement . . . of the Soviet-Chinese state boundary line at some sectors," but the Chinese delegation "attempted to question the state border which had been established historically and consolidated by treaties." The consultations in Peking were not concluded; an agreement was reached to continue them in Moscow on Oct. 15, 1964, but the Chinese side later refused to do so.

According to the Communist Chinese Foreign Ministry, the Soviet Union at the consultation meeting "refused to recognize the treaties relating to the present Sino-Soviet boundary as unequal treaties and obstinately refused to take these treaties as the basis for settling the boundary question between the two countries in its vain attempt to force China to accept a new unequal treaty and thus to perpetuate in legal form its occupation of the Chinese territory which it seized by crossing the boundary line defined by the unequal treaties," Information Department of the Chinese Foreign Ministry, "Chenpao Island Has Always Been Chinese Territory," *PR*, 12:15, no. 11 (Mar. 14, 1969). On Oct. 7, 1969, Communist China announced it had reached agreement with the Soviet Union to resume boundary consultations; "China Announces It Has Reached Accord with USSR to Hold Negotiations on Border Dispute," *New York Times*, Oct. 8, 1969, p. 1.

38. *PR*, 12:6, no. 22 (May 30, 1969).

Communist China has not yet expressly "recognized" any other bilateral treaties concluded by a prior Chinese government, though it has "reconcluded" several treaties covering subjects identical or similar to those covered in earlier treaties.[39] The status of treaties pending reconclusion is vague. In the reconclusion of Kuomintang treaties with the Soviet Union in 1950 and with Afghanistan in 1960 Communist China arranged for the abrogation of earlier treaties by an exchange of notes with its treaty partners.[40] This may indicate that the earlier treaties were not ipso facto terminated by Article 55 of the Common Program. Nevertheless, it is still unclear whether the other contracting parties can invoke those treaties in their relations with Communist China, as the following two cases illustrate.

In 1949, when Communist China took over the United States consular property in Peking, the United States asserted its rights under the 1943 Sino-American Treaty for the Relinquishment of Extraterritorial Rights in China.[41] The Communist Chinese authorities disregarded the legal argument of the United States and continued to occupy the consular property.[42] In the same year, however, Communist China refrained from taking similar measures against Soviet rights in Dairen and Port Arthur; it also did not interfere with Soviet rights to the Chinese Eastern Railway and continued to recognize the independence of Outer Mongolia. All these rights were based on the 1945 Sino-Soviet treaty and accompanying agreements.[43] Nonetheless, a few months later, Communist China

39. *E.g.*, the Treaty of Friendship and Cooperation between Communist China and Czechoslovakia (Mar. 27, 1957), *TYC*, 6:40, in fact covers some subjects similar to those covered by the Sino-Czechoslovakia Treaty of Amity and Commerce (Feb. 12, 1930); *LNTS*, 110:285.

40. See Exchange of Notes Concerning the Abrogation of the Treaty of Friendship and Alliance and of the Agreements on the Chinese Ch'ang Ch'un Railway, Dairen and Port Arthur, Aug. 14, 1945; *UNTS*, 226:16. See also Exchange of Notes on the Abrogation of the "China-Afghanistan Treaty of Friendship," Mar. 2, 1944; *TYC*, 9:14 (1960).

41. 57 Stat. 767; *TS*, no. 984; *UNTS*, 10:261.

42. See "Communists Take U.S. Property in China," *DSB*, 22:119–121, no. 551 (Jan. 23, 1950). It is not clear whether Communist China disregarded the 1943 treaty because it considered it unequal or for some other reason, but an *NCNA* news release called it unequal; "Peking Military Control Committee Requisitions Foreign Barracks in City," *NCNA*, Daily News Release no. 261 (Jan. 19, 1950), p. 77. A relatively recent article also characterizes the treaty as unequal; Shih Hung-ping, "Hou Wai-lu Is An Experienced Anti-Communist," *People's Daily*, Nov. 22, 1966, p. 6.

43. Before Feb. 14, 1950, Communist Chinese sources at least twice reaffirmed the validity of the 1945 Sino-Soviet Treaty and Agreements, once in a North Shensi broadcast on Mar. 18, 1949, and once in a speech by Kuo Mo-jo (then a member of the preparation committee of the Chinese People's Political Consulta-

and the Soviet Union concluded a new treaty in which the Soviet Union agreed to relinquish all rights based on the 1945 Sino-Soviet treaty and accompanying agreements. These two contrasting cases, one involving the United States and the other involving the Soviet Union, suggest that the determination of whether a treaty concluded by a former government continues in force depends largely upon the foreign policy needs of Communist China.

Although the revolutionary change of government in China would justify Communist China's rejection of automatic succession to pre-1949 Chinese treaty obligations, such a principle, in Communist Chinese practice, is not necessarily applicable to similar situations in other states. Thus, after the overthrow of the Nkrumah regime in Ghana on February 24, 1966, the new regime ordered Communist Chinese experts to leave Ghana. Communist China protested[44] that this was a violation of the 1961 Sino-Ghanaian Agreement on Economic and Technical Cooperation,[45] which had been concluded by the Nkrumah regime. Perhaps one can draw a distinction between the Chinese case and the Ghanaian case. In the Communist Chinese view, the Nkrumah regime was "progressive" and the new one is "reactionary"; it may well argue that a reactionary new regime should not have the right to reject treaty obligations contracted by an earlier progressive regime.[46]

In his *International Law*, Oppenheim says that a "vital change of cir-

---

tive Conference) on Aug. 13, 1949; James C. Hsiung, "Communist China's Conception of World Public Order," unpub. diss. (Columbia University, 1967), p. 160.

At a press conference, Sept. 29, 1965, Vice-Premier and Foreign Minister Ch'en Yi observed that "in 1945 Chiang Kai-shek's government concluded a treaty with the government of the Soviet Union recognizing the Mongolian People's Republic. After its founding, New China succeeded to the commitment and recognized Mongolia as a socialist country," in "Vice-Premier Chen Yi's Press Conference: China Is Determined to Make All Necessary Sacrifices for the Defeat of U.S. Imperialism," *PR*, 8:13, no. 41 (Oct. 8, 1965).

44. *E.g.*, see Mar. 6, 1966, note of Communist China to Ghana; "China Strongly Protests against Worsening of Sino-Ghanaian Relations by Ghanaian Authorities," *PR*, 9:8–9, no. 13 (Mar. 25, 1966).

45. *TYC*, 10:250; English trans. in "Text of Sino-Ghanaian Economic and Technical Cooperation Agreement," *SCMP*, 2567:33 (Aug. 28, 1961).

46. According to Kozhevnikov's textbook on international law, when a social revolution remakes a state into a new historical type the new regime need not succeed to all treaty obligations of the former government; see n. 18 above. Thus one may argue that the revolution in Ghana did not remake that state into a new historical type and therefore that the new regime cannot reject any treaty obligation contracted by the former government.

cumstances may be of such a kind as to justify a party in demanding to be released from the obligations of a treaty which cannot be abrogated by unilateral notice."[47] While most modern Western scholars accept this principle, known as the doctrine of *rebus sic stantibus*, its application raises controversial problems. D. P. O'Connell has observed that this is "one of the least satisfactory topics in treaty law."[48]

Concerning this principle, the 1969 Convention on the Law of Treaties provides in Article 62:

> 1. A fundamental change of circumstances which has occurred with regard to those existing at the time of the conclusion of a treaty, and which was not foreseen by the parties, may not be invoked as a ground for terminating or withdrawing from the treaty unless:
> (a) the existence of those circumstances constituted an essential basis of the consent of the parties to be bound by the treaty; and
> (b) the effect of the change is radically to transform the extent of obligations still to be performed under the treaty.
> 2. A fundamental change of circumstances may not be invoked as a ground for terminating or withdrawing from a treaty:
> (a) if the treaty establishes a boundary; or
> (b) if the fundamental change is the result of a breach by the party invoking it either of an obligation of the treaty or of any other international obligation owed to any other party to the treaty.
> 3. If, under the foregoing paragraphs, a party may invoke a fundamental change of circumstances as a ground for terminating or withdrawing from a treaty it may also invoke the change as a ground for suspending the operation of the treaty.[49]

Article 65 of the Convention lays out the procedure to be followed by a contracting party when invoking *rebus sic stantibus*. Except in cases of special urgency, three months' notice of the proposed termination or suspension should be given to the other contracting party. If the other contracting party objects to the action, the parties are to seek a solution through the means indicated in Article 33 of the United Nations Charter: negotiation, enquiry, mediation, conciliation, arbitration, judicial settlement, resort to regional agencies or arrangements, or other peaceful means of their own choice.[50]

47. Oppenheim, *International Law*, I, 939.
48. O'Connell, *International Law*, I, 296. See also the writings cited in Oppenheim, *International Law*, I, 939–944; commentary to art. 59 (fundamental change of circumstances) of Draft Articles on the Law of Treaties adopted by U.N. International Law Commission in 1966, U.N. General Assembly, *Official Records*, 21st sess., supp., no. 9, pp. 85–88 (A/6309/rev.1), 1966.
49. *ILM*, 8:702, no. 4 (July 1969).
50. *Ibid.*, p. 703.

Generally speaking, Communist Chinese writers also accept *rebus sic stantibus* as a principle of treaty law. Chou Keng-sheng, for example, has written: "Although the principle of 'pacta sunt servanda' is a principle of international law, international law, at the same time, also recognizes vital change of circumstances . . . and other conditions as constituting a legitimate reason for the denunciation of a treaty by a contracting party."[51] Apparently he recognizes that, in case of a vital change of circumstances, a state can unilaterally denounce a treaty. This view is not entirely shared by Wang Yao-t'ien, who has argued that when a vital change of circumstances occurs the contracting parties should seek revision or reconclusion of the treaty through diplomatic negotiation. But this principle, he maintains, is not applicable to treaties by which the imperialist states imposed virtual slavery upon other states. In that case, the victim states can unilaterally denounce the treaties.[52]

If the official Communist Chinese position on *rebus sic stantibus* corresponds to Wang's formulation of the principle, there is an important disagreement between Communist China and Article 62 of the 1969 Convention on the Law of Treaties. The provisions of Article 62 specifically exclude application of the principle of *rebus sic stantibus* to treaties establishing boundaries. The commentary to Article 62 prepared by the United Nations International Law Commission explicitly points out that "treaties of cession as well as delimitation treaties" are all within the meaning of treaties establishing boundaries.[53] Many treaties of cession are, however, considered by Communist China or its writers as "unequal treaties" or treaties of a "slavery nature" which – as pointed out by Wang Yao-t'ien – can be unilaterally abolished, regardless of whether there has been a vital change of circumstances.

51. Chou Keng-sheng, "Looking at the West Berlin Question from the Angle of International Law," *KCWTYC*, 1:44 (1959). Another writer, Chou Tze-ya, expresses the same view: "All treaties are concluded under certain circumstances. If the circumstances have already changed, there is no reason for the continued existence of the treaty, especially when the treaty does not specify its duration. The condition, 'conventio omnis intelligitur rebus sic stantibus,' should be implied in such a treaty. Regarding this point, Anglo-French international lawyers all express the same view: If the circumstances under which a treaty was concluded change, and the performance of the treaty would imperil the security of a state, the latter may refuse to perform the treaty obligation, or it at least may demand revision," in "Talks on the Question of Suez Canal," *Hua-tung cheng-fa hsüeh-pao* (East China journal of political science and law), 3:38 (1956).

52. Wang Yao-t'ien, *Kuo-chi mao-yi t'iao-yüeh ho hsieh-ting* (International trade treaties and agreements) (Peking, 1958), p. 15; Wang also criticizes the capitalist countries for frequent invocation of this principle as a pretext to justify unilateral abrogation of treaties.

53. U.N. General Assembly, *Official Records*, 21st sess., supp. no. 9, p. 87 (A/6309/rev.1), 1966.

Because Communist China did not participate in the Conferences adopting the Convention on the Law of Treaties, it is unlikely that it will observe the procedure laid down in Article 65, as it is a *lex ferenda* (the law which it is desired to establish), and not a customary rule of international law, for invoking *rebus sic stantibus* to terminate a treaty. Nevertheless, Communist China's treaty practice does indicate that, under certain circumstances, it will resort to negotiation to settle a dispute on the application of the principle of a vital change of circumstances to a treaty. Following are several examples.

Several Sino-Soviet agreements concerning the Chinese Ch'ang Ch'un Railway, Port Arthur, and Dairen. The Republic of China in 1945 concluded several agreements[54] with the Soviet Union, granting to the latter a military base in Port Arthur, special privileges in the port of Dairen, the right of joint administration and possession of the Chinese Ch'ang Ch'un Railway, and many other privileges. In 1950 the Soviet Union agreed to relinquish all Soviet rights to the Chinese Ch'ang Ch'un Railway not later than 1952, to evacuate its Port Arthur military base, and to renounce its privileges in the port of Dairen. The 1950 agreement referred to the vital change of circumstances in the Far East between 1945 and 1950:

> The Presidium of the Supreme Soviet of the Union of Soviet Socialist Republics and the Central People's Government of the People's Republic of China note that since 1945 the following radical changes have taken place in the situation in the Far East: Imperialist Japan has suffered defeat; the reactionary Kuomintang Government has been overthrown; China has become a People's Democratic Republic; and there has been established in China a new People's Government which has united all China, applied a policy of friendship and co-operation with the Soviet Union, and proved its ability to uphold the State independence and territorial integrity of China and the national honour and dignity of the Chinese people.
>
> The Presidium of the Supreme Soviet of the Union of Soviet Socialist Republics and the Central People's Government of the People's Republic of China consider that because of this new situation a new approach to the question of the Chinese Changchun Railway, Port Arthur and Dairen is possible.
>
> In accordance with these new circumstances the Presidium of the Supreme Soviet of the Union of Soviet Socialist Republics and the

54. Sino-Soviet Agreement on Chinese Ch'ang Ch'un Railway (Aug. 14, 1945), *UNTS*, 10:346; Agreement on Port Arthur (Aug. 14, 1945), *ibid.*, p. 358; Agreement on Dairen (Aug. 14, 1945), *ibid.*, p. 354.

Central People's Government of the People's Republic of China have decided to conclude this Agreement concerning the Chinese Changchun Railway, Port Arthur and Dairen.[55]

The 1888 Convention of Constantinople. During the 1956 Suez Canal crisis, Chou Tze-ya, of the East China Institute of Political Science and Law, argued that the doctrine of *rebus sic stantibus* should be applied to the 1888 Convention of Constantinople.[56] He wrote:

> The Treaty of Constantinople was concluded by the foreign ruler in the name of Egypt with imperialist states when Egypt was ruled by a foreign power [the Ottoman Empire]. There are many provisions infringing on the Egyptian exercise of its sovereignty which are not necessary to guarantee free passage of ships through this interoceanic canal. (For instance, the prohibition in Article 11 against the construction of permanent defense fortifications is an obvious example.) The world situation has changed greatly in the past eighty years. Egypt has already changed from a protected State into an independent state. The existence of this yoke [the Treaty of Constantinople], and the various relationships which have arisen due to this yoke's existence, have certainly imperiled the independence, existence, and development of Egypt. Therefore, Egypt has proposed revision of the Treaty of Constantinople and the convocation of an international conference of all states having ships which pass through the Canal to reconfirm and guarantee the freedom of navigation in the Canal. Such a proposal is obviously necessary in order to show respect for Egyptian sovereignty, and to facilitate international communication.[57]

Several 1945 agreements relating to West Berlin. In its note of November 27, 1958, the government of the Soviet Union stated that the Protocol of the September 12, 1944, Agreement between the governments of the Soviet Union, the United States, and the United Kingdom on the zones of occupation in Germany and on the administration of Greater Berlin,[58] and the related supplementary agreements, including the Agreement on the Control Machinery in Germany concluded between the governments of the Soviet Union, the United States, the United King-

55. Sino-Soviet Agreement Concerning Chinese Ch'ang Ch'un Railway, Port Arthur and Dairen, Feb. 14, 1950; *TYC*, 1:3; *UNTS*, 226:31.

56. Convention Respecting Free Navigation of the Suez Maritime Canal, *BFSP*, 78:18 (1886–1887); *AJIL*, 3:123 *supp.* (1909). Upon obtaining independence, Egypt had accepted the treaty as binding upon it by succession to the Ottoman Empire.

57. Chou Tze-ya, "Talks on the Question of Suez Canal," p. 38.

58. *UST*, 5:2078; *TIAS* 3071.

dom, and France on May 1, 1945[59] – in other words, the agreements that were intended to be in effect during the first years after the capitulation of Germany – had lost their validity and that the occupation regime of West Berlin should be brought to an end.[60] The United States government, in its December 31 reply insisted upon the right of the three Western powers to continue to occupy West Berlin on the basis of the above agreements.[61]

Chou Keng-sheng strongly supported the Soviet position on the Berlin question. His advocacy was partly based on a vital change of circumstances in the German situation:

> Fourteen years after the unconditional surrender of Germany, the development of the German question and the actual situation of present day Germany have fundamentally negated the occupation status of the three Western powers in West Berlin. Germany is now divided into two sovereign states, and the Allies have already formally proclaimed the termination of the state of war with Germany. Between 1954 and 1955, the Soviet Union and the three Western powers respectively terminated their occupation regime in the Democratic Republic of Germany and the Federal Republic of Germany, where they had originally exercised their respective occupation authority. They abolished the occupation laws and regulations and established regular diplomatic relations; the Soviet Union, moreover, formally established diplomatic relations with West Germany. Does it not contradict political realities for the three Western powers to continue to maintain their occupation status in West Berlin in this fundamentally changed political situation? Can the advocate of the theory of formalistic validity of treaties argue for that status?[62]

In Chou's view (p. 45), the Soviet Union could invoke the doctrine of vital change of circumstances to abrogate the several agreements relating to the status of Berlin.

Sino-Soviet Treaty of Friendship, Alliance and Mutual Assistance, February 14, 1950. Article 1 of the Treaty provides: "The two Contracting Parties undertake to carry out jointly all necessary measures within their power to prevent a repetition of aggression and breach of

59. *UNTS*, 236:400; *UST*, 5:2072; *TIAS* 3070.
60. "U.S. Replies to Soviet Note on Berlin," *DSB*, 40:81–89, no. 1021 (Jan. 19, 1959).
61. *Ibid.*, pp. 79–81.
62. Chou Keng-sheng, "Looking at the West Berlin Question," p. 44.

the peace by Japan or any other State which might directly or indirectly join with Japan in acts of aggression."[63] In an interview given by Premier Chou En-lai to a delegation of the Japanese Civilian Radio Union and correspondents of the Kyodo News Agency and the *Asahi Shimbun* on July 25, 1957, the Premier said: "Time and again we have said that after normal relations have been restored, China and Japan can sign a treaty of friendship and non-aggression. In talks with the Delegation of the Japanese Socialist Party, both Chairman Mao Tse-tung and I pointed out that if Japan is to gain full independence, *the provisions of the Sino-Soviet Treaty of Friendship [Alliance and Mutual Assistance] aimed at guarding against the revival of Japanese militarism and its use by others may be revised.*"[64] (Emphasis added.) Chou's statement appears to suggest that a vital change of circumstances may justify the revision of an existing treaty.

Settlement of boundary questions. Communist China has invoked the doctrine of *rebus sic stantibus* in negotiating the settlement of boundary questions with other countries. For example, in his *Report on the Question of the Boundary Line between China and Burma to the Fourth Session of the First National People's Congress* on July 9, 1957, Premier Chou En-lai stated:

> Our government holds that in dealing with the Sino-Burmese boundary question we must adopt a serious attitude toward historical data, [and] we must take a correct stand and viewpoint so as to make a scientific analysis and appraisal of such data and to distinguish between the data which can be used as a legal and reasonable basis and those which have only reference value as a result of changed conditions. At the same time we must bear in mind the fundamental changes of historic importance that have taken place in China and Burma respectively, *i.e.*, China has cast away its semi-colonial status, and both have become independent and mutually friendly countries. The Burmese Government has succeeded to the territory formerly controlled by Britain, and the Union of Burma has been established by combining the various national autonomous states and Burma proper, while our government has taken over the territory under the jurisdiction of the Kuomintang government. In dealing with this boundary question, attention must be paid to these historical changes, and the treaties signed in the past which concern the boundary be-

63. *TYC*, 1:1; *UNTS*, 226:12.
64. "Premier Chou En-lai's Interview with Japanese Press and Radio," *SCMP*, 1582:22 (Aug. 1, 1957).

> tween China and Burma must be treated in accordance with general international practice.[65]

Suspension of cultural cooperation agreements. In accordance with cultural agreements, many students have been exchanged between Communist China and foreign countries.[66] When the "Great Proletarian Cultural Revolution" began in 1966, all Chinese institutions of higher education suspended their classes. On September 20, 1966, the Communist Chinese Ministry of Higher Education notified the Soviet Embassy in Peking that "as the great proletarian cultural revolution is going on in all institutions of higher education in China, classes have been completely suspended, and all foreign students in China are to suspend their studies for one year, and those who are to graduate next summer or by the end of this year may graduate ahead of time."[67] According to Communist China, the Soviet Union did not reply to this note, but instead recalled on its own the Soviet students in China. In the meantime, the Soviet Ministry of Higher and Secondary Special Education on October 7, 1966, informed a representative of the Communist Chinese Embassy in Moscow that:

> Taking into account that in accordance with the agreement on cultural cooperation between the USSR and the CPR of 5 July 1956, and the plan of cultural cooperation between both countries for 1966, the exchange of students, postgraduates, and teachers must be carried out on the basis of reciprocity. The embassy is being informed that the USSR Ministry of Higher and Secondary Special Education and the USSR Academy of Sciences, guided by this principle, have decided to suspend the training of CPR citizens in Soviet educational establishments and research institutions. The Soviet side expects the said Chinese citizens to leave the Soviet Union during October 1966.
>
> It goes without saying that the Soviet side will be ready to consider the question of resuming on the basis of reciprocity the exchange of

65. *WCC*, 4:341, 346, 347; English trans. in Chinese People's Institute, *A Victory for the Five Principles*, pp. 23–24. See also nn. 33–35, 37, 38 above, and accompanying text on Communist China's position toward Sino-Soviet boundary treaties.

66. *E.g.*, art. 2 of the Sino-Soviet Agreement on Cultural Cooperation provides: "The Contracting Parties have decided . . . 4. To exchange visits of students, graduate students and teachers, who wish to improve their qualifications," *TYC*, 5:152; *UNTS*, 263:137.

67. Quoted from Chinese note to Soviet Embassy in China, Oct. 22, 1966; "Chinese Foreign Ministry Lodges Strongest Protest Against Soviet Authorities for Unjustifiably Suspending Studies of All Chinese Students in USSR," *SCMP*, 3809: 39 (Oct. 27, 1966).

students, postgraduates, and trainees as soon as the Chinese side displays a readiness to resume such exchanges.[68]

The Soviet decision to suspend the studies of all Chinese students in the Soviet Union drew a protest from Communist China. On October 22, 1966, the Communist Chinese Foreign Ministry delivered a note of protest to the Soviet Embassy in Peking that stated in part:

> The pretext of the Soviet authorities is utterly untenable. The whole world knows that a great cultural revolution is in progress in China and the suspension of studies for one year does not apply to the Soviet students alone but to all foreign students in China. Since in the Soviet Union the case does not exist in which all foreign students are required to suspend their studies for one year, the unilateral decision to suspend the studies of the Chinese students there is clearly a discrimination against the Chinese students exclusively.[69]

The note seems to suggest that the cultural revolution in China was a vital change of circumstances which justified its suspension of that part of the cultural cooperation agreement concerning foreign students; since no similar vital change of circumstances existed in the Soviet Union, suspension by the latter of Chinese students was unjustified.

Albanian withdrawal from the 1955 Warsaw Treaty. On May 14, 1955, the Soviet bloc countries signed a treaty of friendship, cooperation, and mutual assistance at Warsaw.[70] The contracting parties included Poland, East Germany, Czechoslovakia, Bulgaria, the Soviet Union, Hungary, Rumania, and Albania. The Warsaw Treaty took effect on June 6, 1955, in accordance with Article 10 of the treaty. Article 11 provided that it would "remain in force for twenty years." There was no provision for withdrawal from the treaty before expiration of the twenty-year period.

On September 13, 1969 – before expiration of the twenty-year period – Albania announced its unilateral withdrawal. One of the reasons cited by the Albanian People's Assembly referred to changed circumstances:

> The conclusion of the Warsaw Treaty was aimed at guaranteeing with joint forces the security of the participating states from any imperialist aggression, especially from the North Atlantic bloc headed by U.S.

68. "Ceremonies in Peking – Exchange Students Expelled," *CDSP*, 18:14, no. 40 (Oct. 26, 1966).
69. See n. 67 above.
70. *UNTS*, 219:3.

> imperialism, as well as at strengthening all-round cooperation among the socialist states on the basis of the lofty principles of Marxism-Leninism and proletarian internationalism. However, the Soviet Government and the governments of some other member states of the Warsaw Treaty, in opposition to the fundamental provisions of the treaty, have thoroughly distorted by their actions the aims for which the treaty was created. All their activities proceed from the imperialist aims of the Soviet-U.S. collaboration for the domination of the world to the detriment of the fundamental interests of the peoples.[71]

The Albanian denunciation mentioned neither the twenty-year period nor any principle of international law. However, the above quotation uses language that sounds very much like *rebus sic stantibus.*

Communist China in several official statements admired and "resolutely" supported the Albanian action.[72] None of the statements invoked international law to support the Albanian action. But all of them did refer to the aggressive nature of the Warsaw Treaty, implying that the nature of that treaty had changed. Communist China's position seems to suggest that a change of circumstances transforming the essential nature of a treaty would justify a contracting state to terminate the treaty by unilateral action.

Oppenheim believes that violation of a treaty by one of the contracting parties gives the other party the discretion to cancel the treaty on that ground.[73] Communist Chinese writers seem to agree; for example, Chou Keng-sheng has written that "international law . . . recognizes . . . violation of a treaty by the other contracting party . . . as constituting [a] legitimate reason for a contracting party to denounce the treaty."[74]

During the 1958 Berlin crisis, Communist China supported the Soviet proposal for an end to occupation of the city. In a statement issued on December 21, 1958, it maintained that:

> Fostered energetically by [the United Kingdom, France, and the United States], particularly by the U.S. imperialists, militarism has

71. "Albanian People's Assembly Passes Law Denouncing Warsaw Treaty, Announces Albania's Withdrawal from It," *PR*, 11:7, no. 38 (Sept. 20, 1968).

72. See "Courageous and Resolute Revolutionary Action" (editorial), *People's Daily*, Sept. 20, 1968, p. 1, English trans. in *PR*, 11:9–10, no. 39 (Sept. 27, 1968); Sino-Albanian Press Communiqué, Peking, Nov. 20, 1968, *PR*, 11:20–21, no. 48 (Nov. 29, 1968); and "At Tirana Mass Rally: Speech [Dec. 2, 1968] by Comrade Huang Yung-sheng (Excerpt)," *PR*, 11:7–9, no. 50 (Dec. 13, 1968).

73. Oppenheim, *International Law*, I, 8th ed. (London, 1955), 947.

74. Chou Keng-sheng, "Looking at the West Berlin Question," p. 44. See also Wang Yao-t'ien, *International Trade Treaties*, p. 15.

> revived in West Germany and West Berlin. They have further mis-used the occupation regime in Berlin to turn West Berlin into a "front line city" within Democratic Germany, a base for conducting the cold war, creating tension, and undertaking subversive and disruptive activities against Democratic Germany and other socialist countries. All this proves that the United States, Britain, and France have long since thoroughly violated the fundamental principles of the Potsdam Agreement and, therefore, have forfeited all legal grounds for their continued occupation of West Berlin.[75]

Some Western writers have suggested that only a violation of the essential terms of a treaty creates a right of cancellation in the other contracting party.[76] The 1969 Convention on the Law of Treaties adopts a similar view. Article 60, paragraph 1, of the Convention provides that "a *material* breach of a bilateral treaty by one of the parties entitled the other to invoke the breach as a ground for terminating the treaty or suspending its operation in whole or in part."[77] (Emphasis added.)

In Communist Chinese practice there seems to be a tendency to protest against the alleged violation of treaties by the other party, but not to take the initiative to abrogate the treaty on that ground. To illustrate, the Communist side repeatedly violated subparagraph (d) of the Armistice Agreement of 1953 by introducing weapons into Korea. As a result, on June 21, 1957, the United Nations Command announced to the Communist side that in view of these violations it intended to abrogate that part of the Armistice Agreement relating to the introduction of arms.[78] On June 27, 1957, Communist China protested to the United Nations Command and declared that such an act constituted a "flagrant violation of the Korean Armistice Agreement."[79] It did not, however, choose to abrogate the Agreement.

A second illustration relates to the suspension of study of foreign students in China for the purposes of carrying out the Great Proletarian Cultural Revolution. Czechoslovakia, Bulgaria, and Hungary retaliated by expelling Chinese students in their countries or refusing to readmit

75. *WCC*, 5:219, 220; English trans. in "Chinese Government Supports Soviet Proposal on Berlin Question," *SCMP*, 1921:46 (Dec. 24, 1958).

76. See Oppenheim, above, n. 73.

77. *ILM*, 8:701, no. 4 (July 1969).

78. Report Concerning Action Taken by the United Nations Command to Maintain Military Balance in Korea: Special Report by the Unified Command in Korea Transmitted to the U.N. Secretary-General by the U.S. Representative, Aug. 9, 1957; U.N. Doc. A/3631 (1957).

79. *WCC*, 4:340; English trans. in "China Protests at U.S. Action in South Korea," *SCMP*, 1562:29–30 (July 3, 1957).

those who had returned to China to participate in the revolution. Communist China protested against those measures on the ground that they violate cultural agreements between Communist China and those countries.[80] But Communist China has not formally abrogated the agreements on that ground.

In its treaty practice, Communist China does not seem to have formally abrogated any treaty on the ground of violation of the treaty by the other party. Even in the case of serious violations, it has chosen to suspend rather than abrogate. When Burmese mobs in 1967 attacked Chinese experts sent to Burma under the Sino-Burmese Agreement on Economic and Technical Cooperation,[81] Communist China merely protested to the Burmese government and suspended the Agreement despite its charge that the mob attacks had been instigated by the Burmese government.[82]

In its various statements on the Vietnamese question, Communist China has repeatedly charged the United States and the Republic of Vietnam with "violations" of the 1954 Geneva Agreements on Indo-

80. See Communist Chinese Foreign Ministry's Aide-Memoire to the Hungarian Embassy in China, Nov. 17, 1966, "China Protests at Hungary's Not Allowing Chinese Students to Continue Their Studies," *SCMP*, 3826:30–31 (Nov. 23, 1966); Foreign Ministry's Note to the Bulgarian Embassy in China, June 10, 1967, "Chinese Foreign Ministry Sends Protest Note to Bulgarian Embassy," *SCMP*, 3959:38–39 (June 14, 1967); Foreign Ministry's Note to Czechoslovakian Embassy in China, July 27, 1967, "Protest Against Czechoslovak Revisionist Clique's Breach of Cultural Cooperation Agreement," *PR*, 10:54, no. 32 (Aug. 4, 1967).

81. Signed Jan. 9, 1961; *TYC*, 10:369; English trans. in "Sino-Burmese Agreement on Economic and Technical Cooperation," 4 *PR*, 4:8–9, no. 2 (Jan. 13, 1961).

82. Note delivered by Communist Chinese Embassy in Rangoon to Burmese Foreign Ministry, July 5, 1967, "Chinese Embassy Protests against Burmese Government's Sabotage of Economic-Technical Cooperation Agreement," *SCMP*, 3977:27–28 (July 11, 1967). Conclusion of the note: "This series of fascist atrocities committed by the reactionary Burmese Government have seriously sabotaged the implementation of the Sino-Burmese agreement on economic and technical cooperation and directly endangered the personal security of the Chinese experts working in Burma, making it very difficult for them to continue working normally. The embassy of the People's Republic of China deems it necessary to warn the reactionary Burmese Government sternly that it should bear full responsibilities for all the serious consequences arising from its anti-China crimes."

On Oct. 31, 1967, the Communist Chinese government issued a statement that the Sino-Burmese Agreement on Economic and Technical Cooperation "ha[d] been torn to pieces" by the Burmese government because of the latter's expressed decision to expel Chinese experts from Burma (note of Oct. 6, 1967); "Chinese Government Statement: Most Strongly Protesting against the Ne Win Government's Expulsion of Chinese Experts," *PR*, 10:27–28, no. 46 (Nov. 10, 1967). It is not clear whether the phrase "ha[d] been torn to pieces" has the same implication as "abrogation."

china[83] in spite of the fact that neither of them signed or acceded to the Agreements. When the Geneva Agreements were signed, in 1954, the United States' representative made the following statement:

> [The United States] will refrain from the threat or the use of force to disturb them, in accordance with Article 2 (4) of the Charter of the United Nations dealing with the obligation of members to refrain in their international relations from the threat or use of force; and . . . it would view any renewal of the aggression in violation of the aforesaid agreements with grave concern and as seriously threatening international peace and security . . .
>
> With respect to the statement made by the representative of the State of Viet-Nam, the United States reiterates its traditional position that peoples are entitled to determine their own future and that it will not join in an arrangement which would hinder this. Nothing in its declaration just made is intended to or does indicate any departure from this traditional position.[84]

The state of Vietnam[85] (now Republic of Vietnam) also made a statement, which said in part: "The Government of the State of Viet-Nam requests that note be made of its solemn protest against the manner in which the armistice has been concluded and against the conditions of the armistice which take no account of the profound aspirations of the Vietnamese people, and of the fact that it reserves to itself complete freedom of action to guarantee the sacred right of the Vietnamese people to territorial unity, national independence and freedom."[86]

These statements can hardly be interpreted as accessions to the 1954 Geneva Agreements. Communist China's disregard of such legal niceties and its repeated charges stand, all the same. Chung Ho, for instance, has

83. The Geneva Agreements include: (1) Final Declaration on Indochina; (2) Agreements on the Cessation of Hostilities between the Franco-Vietnamese Command and the Command of the People's Army of Vietnam; (3) Agreements on the Cessation of Hostilities between Franco-Laotian Command and the Command of the People's Army of Vietnam; (4) Agreement on the Cessation of Hostilities between the Royal Khmer Army Command and the Command of the People's Army of Vietnam; see *DIA* (1954), pp. 138–143.

84. *American Foreign Policy, 1950–1955, Basic Documents,* I (Washington, 1957), 787, 788.

85. By a referendum held Oct. 23, 1955, the State of Vietnam abolished its monarchical system and adopted the republican system; since then it has been renamed the Republic of Vietnam; "Results of Referendum in Viet-Nam," *DSB*, 33: 760, no. 854 (Nov. 7, 1955).

86. Protest by the Vietnamese Delegation against the Geneva Conference Agreements, July 21, 1954; *Documents on American Foreign Relations, 1954* (Washington, D.C., 1955), pp. 315–316.

made a long list of acts of the United States and the Republic of Vietnam which allegedly violate various articles of the Agreements.[87]

In June 1966 the United States began to bomb Hanoi. On July 3, 1966, Communist China made a statement which implied that, as a result of the United States' violation of the Geneva Agreements, it would no longer be bound by those agreements:

> U.S. imperialism long ago completely violated the Geneva agreements and broke the line of demarcation between southern and northern Vietnam. It has now further broken this line by its bombing of the capital of the heroic Vietnamese people. The United States must be held responsible for all the serious consequences arising therefrom.
>
> With the breaking of the line of demarcation by the United States, the Vietnamese people have ceased to be subject to any restrictions. All the countries and people that genuinely support the Vietnamese people's war of resistance against U.S. aggression have also ceased to be subject to any restrictions.[88]

Recently, an article by "Commentator" in *People's Daily* said that "the Geneva agreements were long ago torn to shreds by U.S. Imperialism and not a trace remains."[89] The above statements suggest that Communist China feels released from its obligations under the Geneva Agreements because of violations of the agreements by third parties – the United States and the Republic of Vietnam. This seems to be a new principle of international law developed by Communist China that can hardly find support in Western international law treatises.[90]

In a discussion of the West Berlin question, Chou Keng-sheng has written that international law recognizes "the disappearance of the purpose

87. Chung Ho, "How U.S. Imperialism Has Torn Up the Geneva Agreements," *PR*, 8:24, no. 14 (Apr. 2, 1965).

But some Western writers feel that some unilateral declarations issued by a state may be legally binding; *e.g.*, Brownlie, *Principles of Public International Law*, (London, 1966), pp. 511–512. If so, Communist China could possibly invoke this rule to charge the U.S. with violating its unilateral pledge (1954) to respect the Geneva Agreements. Communist China has not yet explicitly made such a charge against the U.S.

88. "Chinese Government Statement: China's Aid to Vietnam in Fighting U.S. Aggression Further Ceases to Be Subject to Any Bounds or Restrictions," *PR*, 9:19, no. 28 (July 8, 1966).

89. "Insidious U.S.-Soviet Collaboration Policy," *People's Daily*, July 23, 1967, p. 4; English trans. in *PR*, 10:13, no. 31 (July 28, 1967).

90. Communist China might invoke the doctrine of *rebus sic stantibus* to support its position, arguing that measures taken by the U.S. and the Republic of Vietnam have fundamentally changed the Indochina situation, justifying suspension or abrogation of the Geneva Agreements.

of a treaty . . . as constituting a legitimate reason for a contracting party to denounce the treaty." In his view, the West Berlin situation would call for application of this principle:

> According to Article 4 of the Convention on Relations between the United Kingdom, the United States, and France and the Federal Republic of Germany, signed at Bonn in 1952 and amended by the Paris Conference in 1954 (referred to as the Bonn Treaty) . . . the stationing of forces in the West Berlin area by the three Western powers is [for the defense of the free world and is therefore] completely divorced from its relationship with the original agreement [concerning the occupation zones concluded in 1945], the purpose of which was to occupy and control Germany . . . In essence, the three Western powers, by remaining in West Berlin, treat Berlin as a base for engaging in hostile and subversive activities against the socialist countries for the North Atlantic Treaty aggressive bloc. It is inconceivable that one of the Allies – the Soviet Union – who signed the agreement concerning the occupation zones, is still obligated to tolerate the other three powers' use of the occupation rights prescribed in this obsolete and meaningless agreement to carry out threats against her and other socialist countries.[91]

Apparently, the principle referred to by Chou is derived from his reading of a section of Oppenheim's *International Law* which states: "All treaties the obligations of which concern a certain object become void through the extinction of such object, for example, treaties concluded in regard to a third state when such State merges in another."[92] It is clear that the term "object" as used in Oppenheim refers to the "subject matter" rather than the "purpose" of a treaty. Interestingly, however, both "object" and "purpose" may be translated by the same Chinese term, *mu-ti*; and in fact several Chinese writers before 1949 did use *mu-ti* to translate "object" from the earlier edition of the above passage from Oppenheim.[93] But the correct translation of "object" as used in this context should be, as Ts'ui Shu-ch'in recognizes, *k'o-t'i*.[94] Chou's unjustified invocation of Oppenheim's principle apparently was caused by his confusion over the two different English terms translated by the same Chinese term.

91. Chou Keng-sheng, "Looking at the West Berlin Question," p. 44.
92. Oppenheim, *International Law*, I, 945.
93. *E.g.*, Wang Fu-yen, *Kuo-chi kung-fa lun* (On the public international law), I, (Shanghai, 1933), 481–482.
94. Ts'ui Shu-ch'in, *Kuo-chi-fa* (International law), I, (Shanghai, 1947), 239.

The establishment of formal diplomatic relations is not a prerequisite for concluding a treaty or agreement between Communist China and other states. Thus, the lack of diplomatic relations has not prevented the conclusion of agreements between Communist China and other countries such as the United States[95] and Lebanon.[96]

As to the question of the effect of severance of diplomatic relations upon existing treaty relations between contracting parties, Article 63 of the 1969 Vienna Convention on the Law of Treaties provides: "The severance of diplomatic or consular relations between parties to a treaty does not affect the legal relations established between them by the treaty except in so far as the existence of diplomatic or consular relations is indispensable for the application of the treaty."[97] The Convention is silent on the question of the effect of suspension of diplomatic or consular relations upon treaty relations between contracting states. However, a fortiori, it appears that Article 63 of the Convention should also be appliable to the case of suspension; that is to say, the suspension of diplomatic or consular relations between contracting states does not automatically terminate their treaty relations unless the existence of such relations "is indispensable for the application of the treaty." No Communist Chinese writer has ever discussed these two questions. Moreover, the actual practice of the nation is not clear.

Communist China has so far not formally severed diplomatic relations with any state, though it has suspended them with several (Indonesia, Burundi, the Central African Republic, Dahomey, and Ghana). Because none of these countries has yet "resumed" diplomatic relations with Communist China, it is not clear whether the "suspension" of diplomatic relations terminates or merely suspends treaty relations.

The Republic of Indonesia does not seem to regard the suspension of relations as having the effect to automatically suspending or terminating the 1955 Sino-Indonesian Dual Nationality Treaty.[98] After the suspension (in October 1966), many Chinese in Indonesia were granted Indonesian citizenship through their exercise of the right of option under the procedure provided in Article 6 of the treaty, until the Indonesian De-

95. Sino-American Agreed Announcement on the Return of Civilians, Sept. 10, 1955, *TYC*, 4:1; "U.S., Red China Announce Measures for Return of Civilians," *DSB*, 33:456, no. 847 (Sept. 19, 1955).
96. Sino-Lebanese Trade Agreement, Dec. 31, 1955; *TYC*, 4:159.
97. *ILM*, 8:703, no. 4 (July 1969).
98. *TYC*, 8:12.

partment of Justice finally suspended the procedure.[99] On April 10, 1969, the Republic of Indonesia unilaterally abrogated the Dual Nationality Treaty – a move that, under the terms of the treaty itself, is difficult to justify.[100] Article 14 of the treaty provides its validity for twenty years and its enforcement thereafter unless one of the parties desires to terminate it. The Indonesian government justified its unilateral abrogation on the ground that "a fundamental change in circumstances" had occurred in the relations between Indonesia and Communist China.[101]

According to information contained in the *People's Daily* up to September 30, 1969, Communist China has not yet officially reacted to this abrogation. But while the bill on abrogation of the treaty was pending in the Indonesian Parliament, an *NCNA* news commentary denounced the Indonesian government for "turn[ing] a blind eye on the stipulations written in black and white in the treaty between the two countries, and tr[ying] to annul it unilaterally."[102]

The effect of war upon treaties between belligerents is a controversial question among Western international law scholars.[103] Oppenheim sets forth the following principles:

> The outbreak of war cancels all political treaties between the belligerents (as, for instance, treaties of alliance) which have not been concluded for the purpose of setting up a permanent condition of things . . .
>
> Political and other treaties concluded for the purpose of setting up a permanent condition of things . . . are not *ipso facto* annulled by the outbreak of war; but nothing prevents the victorious party from imposing by the treaty of peace alteration in, or even the dissolution of, such treaties.[104]

Communist Chinese scholars have challenged these rules by arguing that war automatically cancels *all* treaties between belligerents, including

99. See Mochtar Kusuma-atmadja, "The Sino-Indonesian Dual Nationality Treaty," in Cohen, ed., *China's Practice of International Law* (to be pub. 1971), pp. (ms.) 36–37.

100. *Ibid.*

101. See also *ibid.*, pp. 37–38.

102. "Indonesian Fascist Regime Attempts to Tear Up Sino-Indonesian Dual Nationality Treaty," *SCMP*, 4302:34 (Nov. 20, 1968).

103. See McNair, *Legal Effects of War* (Cambridge, Eng., 1948).

104. Oppenheim, *International Law*, II, 7th ed. (London, 1952), 303–304.

treaties concluded for the purpose of setting up a permanent condition of things, such as treaties ceding a piece of territory. Thus, Shao Chin-fu writes:

> After the Sino-Japanese War of 1894, the government of the Ch'ing Dynasty, by signing the Treaty of Shimonoseki, ceded Taiwan and Penghu to Japan. With the outbreak of China's War of Resistance against Japan in 1937, *in accordance with international law, the treaties between the countries became null and void. The Treaty of Shimonoseki was no exception.* In 1945, after China's victory in the Anti-Japanese War, China recovered these two places from Japan . . . Since Taiwan has always been Chinese territory, it is a matter of course for China to take it back like a thing restored to its original owner. *It is not a case of China taking a new territory from Japan which must be affirmed by a peace treaty.* [Emphasis added.] [105]

Western scholars generally hold the view that Taiwan, although part of Chinese territory since the seventeenth century, became de jure Japanese territory as a result of the 1895 Treaty of Shimonoseki. In their view – politically – Taiwan was "restored" to China after the Japanese surrender in 1945, but legalistically speaking China has acquired a new territory from Japan and its acquisition should be regulated by a peace treaty or by other principles of international law.[106]

Communist Chinese scholars, however, have considered the 1895 Treaty of Shimonoseki an unjust treaty.[107] In their view, its abrogation as a result of the outbreak of war automatically transferred Taiwan to China. They insist that the usual rules relating to acquisition of new territory do not apply to the righting of a historical wrong.[108]

105. "The Absurd Theory of 'Two Chinas' and Principles of International Law," *KCWTYC*, 2:14 (1959). See also Mei Ju-ao, "Stripping the Aggressor of Its Legal Pretext," *People's Daily*, Jan. 31, 1955, p. 3; English trans. entitled "The Aggressor and the Law," *PC*, 5:10 (Mar. 1, 1955).

106. See chap. ii above, nn. 29–39.

107. Shih Sung *et al.* have argued: "It is clearly recognized in modern international law that in order to correct historical inequities, sovereignty over national territory must be restored," in "An Initial Investigation into the Old Law Viewpoint in the Teaching of International Law," *CHYYC*, 4:14, 16 (1958).

108. The Republic of China's position on the effect of the Sino-Japanese War on the 1895 Treaty of Shimonoseki is similar. On Oct. 25, 1950, President Chiang Kai-shek said: "In 1941, our Government declared war on Japan, thus automatically abolishing the Treaty of Shimonoseki and legally restoring Taiwan and Penghu to China . . . With the victorious conclusion of World War II, Japan accepted this [Cairo] declaration and China took over Taiwan and Penghu. The people of Taiwan are Chinese, and there is no doubt that the sovereignty of Taiwan belongs to the Republic of China, racially, historically, culturally, legally," Office of Government

On the whole, Communist China's treaty law and practice are not widely different from those of the West. But, where differences exist, they are profound. This is especially apparent in the problem of unequal treaties and in the perceived effect of revolutionary change of government upon the treaty obligations of a state. To what extent these differences are caused by Communist ideology is difficult to say; Communist Chinese writings do not deal with this question. Yet the very fact that Communist ideology does not specifically comment on these problems may indicate that policy needs rather than philosophical perceptions dictate the distinctive approaches taken by Communist China to some treaty problems.

There is no evidence that traditional Chinese concepts of foreign relations, such as the hierarchical tributary system of world order, have any influence on Communist China's treaty law and practice. In fact, the form and content of treaties concluded by Communist China do not reveal any especially "Chinese" characteristics. Moreover, most of the principles and rules of treaty law invoked by Communist China or its writers can also be found in Soviet and Western treaty law and practice.

This does not mean that China's past experience has no connection with Communist China's law and practice of treaties. On the problem of unequal treaties, for instance, the influence of China's experience since the Opium War of 1839–1842 has been profound. The unequal treaties have been denounced not only by the Communist Chinese government, but also by the Nationalist government and the pre-1928 Peking government. The unequal treaty concept is developed primarily from China's past experience.

Throughout the 1950's, Soviet influence over Communist China's treaty law and practice was relatively significant. The institution of ratification and approval of treaties, the emphasis on treaties as a principal source of international law, and the unilateral abrogation by the revolutionary government of treaties made by the previous regime are several instances where the Soviet influence is particularly observable. Communist Chinese writers, however, have cited Western more often than Soviet

Spokesman (ed.), *Selected Speeches and Messages of President Chiang Kai-shek, 1949–52* (Taipei, 1952), p. 65. Note that the Treaty of Peace between the Republic of China and Japan of Apr. 28, 1952, confirmed, inter alia, abrogation of the 1895 Treaty of Shimonoseki. Art. 4 of the Treaty of Peace: "It is recognized that all treaties, conventions and agreements concluded before December 9, 1941, between China and Japan have become null and void as a consequence of the war," *UNTS*, 136:45.

authorities for support on some problems in the law of treaties. Moreover, Communist China's diplomatic notes usually refer to "international practice," rather than to the specific rules developed by the Soviet Union and other Communist states.

The fact that a principle of treaty law is accepted both by Communist China and the West does not guarantee identical application of that principle. In applying a principle of treaty law, political considerations are Communist China's major concern. Of course, the difference between Communist China and the West in this respect is a matter of degree. Nevertheless, its treaty practice indicates that, in comparison with Western countries, Communist China has shown less concern for purely legal principles than for political factors.

Appendix A: Note on the Study of the Law of Treaties in Communist China

In the Republican period (and now in the Republic of China on Taiwan), the law of treaties was one of the areas of international law which received considerable attention from Chinese scholars. Many books and articles discussing various problems of the law of treaties were published during this period.[1] In universities or colleges a course on it has usually been offered in the department of political science or in the department of diplomacy.

Since the establishment of the Communist Chinese regime in 1949, there has been a steady decline in study of the law of treaties. Writers in Communist China rarely discuss the multifarious problems of the subject, such as reservation, ratification, and termination. Nor do they seem to pay any attention to the recent development of the law of treaties either in the Soviet Union[2] or in non-Communist states. Thus, for instance, no Communist Chinese writers refer to the various drafts on the law of treaties considered by the U.N. International Law Commission that have been widely discussed by many scholars. Moreover, it does not appear that a course on the law of treaties has been offered in any of the Communist Chinese universities, colleges, or institutes. On the other hand, in their discussion of some practical problems of international law (such as the Sino-Indian border dispute and the West Berlin question), some Communist Chinese writers have dealt with certain of its principles or rules.

As a result of the Cultural Revolution, university education in Communist China was suspended for 1966–67. In mid-1967 Communist China tried to reactivate the colleges and universities, but apparently they have not functioned normally since reopening.

Since late 1965, professional writers seem to have ceased writing arti-

1. To mention just a few: Wu K'un-wu, *T'iao-yüeh lun* (On treaties) (Shanghai, 1933); Huang Teng Young, *The Doctrine of Rebus Sic Stantibus in International Law* (Shanghai, 1935); Tseng Yu-hao, *The Termination of Unequal Treaties in International Law* (Shanghai, 1933); Wang T'ieh-ya, *Chan-cheng yü t'iao-yüeh* (War and treaties) (Chungking, 1944).

2. In the 1950's Communist Chinese writers translated a few Soviet articles on the subject; following is a list of articles I know of. Kozhevnikov, "Nekotorye voprosy teorii i praktiki mezhdunarodnogo dogovora," *Sovetskoe gosudarstvo i pravo*, 2:62 (1954), trans. in *Kuo-chi wen-t'i yi-ts'ung* (Collected translations of international problems), 11:104 (1954); Tunkin, "Parishskie Soglasheniia i mezhdunarodnoe pravo," *Sovetskoe gosudarstvo i pravo*, 2:13 (1955), trans. in *Kuo-chi wen-t'i yi-ts'ung*, 9:77 (1955); Durdenevskii and Krylov, "Narushenie imperialisticheskimi gosudarstvami printsipa sobliudeniia mezhdunarodnykh dogovorov," *Mezhdunarodnaia zhizn'*, 2:64 (1955), trans. in *Kuo-chi wen-t'i yi-ts'ung*, 6:87 (1955); Tunkin, "O nekotorykh voprosakh mezhdunarodnogo dogovora v sviazi s Varshavskim dogovorom," *Sovetskoe gosudarstvo i pravo*, 1:98 (1956), trans. in *Kuo-chi wen-t'i yi-ts'ung*, 5:88 (1956).

In 1959 Kozhevnikov's *Mezhdunarodnoe pravo* (Moscow, 1957), which contains a chapter on international agreements, was translated into Chinese; see pp. 242–276, trans. in Su-lien k'o-hsüeh-yüan fa-lü yen-chiu-so pien, *Kuo-chi fa* (International law), 248–284 (Peking, 1959).

cles on international law, though the *People's Daily* has occasionally published a few somewhat relevant articles. Generally, the articles have been written by commentators of the *People's Daily*, by Red Guard elements, or by members of the Chinese People's Liberation Army. It is possible that some Communist Chinese writers on international law were purged or their views denounced during the Cultural Revolution, but no information concerning this point has been available to me.

Appendix B: Treaty of Friendship, Alliance and Mutual Assistance between the Union of Soviet Socialist Republics and the People's Republic of China*

(February 14, 1950)

The Presidium of the Supreme Soviet of the Union of Soviet Socialist Republics and the Central People's Government of the People's Republic of China,

Being determined, by strengthening friendship and co-operation between the Union of Soviet Socialist Republics and the People's Republic of China, jointly to prevent the revival of Japanese imperialism and the repetition of aggression on the part of Japan or of any other State that might in any way join with Japan in acts of aggression,

Being anxious to promote a lasting peace and general security in the Far East and throughout the world in accordance with the purposes and principles of the United Nations,

Being firmly convinced that the strengthening of good-neighbourly and friendly relations between the Union of Soviet Socialist Republics and the People's Republic of China is in accordance with the fundamental interests of the peoples of the Soviet Union and China,

Have decided for this purpose to conclude the present Treaty and have appointed as their plenipotentiaries:

The Presidium of the Supreme Soviet of the Union of Soviet Socialist Republics: Andrei Yanuarevich Vyshinsky, Minister of Foreign Affairs of the USSR;

The Central People's Government of the People's Republic of China: Chou En-lai, Chairman of the State Administrative Council and Minister of Foreign Affairs of China.

The two plenipotentiary representatives, having exchanged their full powers, found in good and due form, have agreed as follows:

ARTICLE 1

The two Contracting Parties undertake to carry out jointly all necessary measures within their power to prevent a repetition of aggression and breach of the peace by Japan or any other State which might directly or indirectly join with Japan in acts of aggression. Should either of the Contracting Parties be attacked by Japan or by States allied with Japan and thus find itself in a state of war, the other Contracting Party shall immediately extend military and other assistance with all the means at its disposal.

The Contracting Parties likewise declare that they are prepared to participate, in a spirit of sincere co-operation, in all international action designed to safeguard peace and security throughout the world, and will devote all their energies to the speediest realization of these aims.

*_TYC_, 1:1–3; English trans. in _UNTS_, 226:12, 14, 16.

ARTICLE 2

The two Contracting Parties undertake, by common agreement, to strive for the conclusion at the earliest possible date, in conjunction with the other Powers which were their Allies during the Second World War, of a Peace Treaty with Japan.

ARTICLE 3

Neither of the Contracting Parties shall enter into any alliance directed against the other Party, or participate in any coalition or in any action or measures directed against the other Party.

ARTICLE 4

The two Contracting Parties shall consult together on all important international questions involving the common interests of the Soviet Union and China, with a view to strengthening peace and universal security.

ARTICLE 5

The two Contracting Parties undertake, in a spirit of friendship and cooperation and in accordance with the principles of equal rights, mutual interests, mutual respect for State sovereignty and territorial integrity, and non-intervention in the domestic affairs of the other Party, to develop and strengthen the economic and cultural ties between the Soviet Union and China, to render each other all possible economic assistance and to effect the necessary economic co-operation.

ARTICLE 6

This Treaty shall come into force immediately upon ratification;† the exchange of the instruments of ratification shall take place at Peking.

This Treaty shall remain in force for thirty years. If neither of the Contracting Parties gives notice one year before the expiration of the said period that it wishes to denounce the Treaty, it shall remain in force for a further five years and shall thereafter be continued in force in accordance with this provision.

DONE at Moscow, on 14 February 1950, in two copies, each in the Russian and Chinese languages, both texts being equally authentic.

By authorization
of the Presidium of the Supreme Soviet
of the Union
of Soviet Socialist Republics:
(Signed) A.Y. Vyshinsky

By authorization
of the Central People's
Government of the People's
Republic of China:
(Signed) Chou En-lai

†Came into force Apr. 11, 1950.

Appendix C: Agreement between the Government of the People's Republic of China and His Majesty's Government of Nepal on the Question of the Boundary Between the Two Countries*

(March 21, 1960)

The Government of the People's Republic of China and His Majesty's Government of Nepal have noted with satisfaction that the two countries have always respected the existing traditional customary boundary line and lived in amity. With a view to bringing about the formal settlement of some existing discrepancies in the boundary line between the two countries and the scientific delineation and formal demarcation of the whole boundary line, and to consolidating and further developing friendly relations between the two countries, the two Governments have decided to conclude the present Agreement under the guidance of the Five Principles of Peaceful Coexistence and have agreed upon the following:

ARTICLE I

The Contracting Parties have agreed that the entire boundary between the two countries shall be scientifically delineated and formally demarcated through friendly consultations, on the basis of the existing traditional customary line.

ARTICLE II

In order to determine the specific alignment of the boundary line and to enable the fixing of the boundary between the two countries in legal form, the Contracting Parties have decided to set up a joint committee composed of an equal number of delegates from each side and enjoin the committee, in accordance with the provisions of Article III of the present Agreement, to discuss and solve the concrete problems concerning the Sino-Nepalese boundary, conduct survey of the boundary, erect boundary markers, and draft a Sino-Nepalese boundary treaty. The joint committee will hold its meetings in the capitals or other places of China and Nepal.

ARTICLE III

Having studied the delineation of the boundary line between the two countries as shown on the maps mutually exchanged (for the map submitted by the Chinese side, see attached Map 1; for the map submitted by the Nepalese side, see attached Map 2)† and the information furnished by each side about its actual jurisdiction over the area bordering on the other country, the Contracting Parties deem that, except for discrepancies in certain sections, their understanding of the traditional customary

*TYC, 9:63–65; English trans. in Chinese People's Institute of Foreign Affairs, *New Development in Friendly Relations between China and Nepal* (Peking, 1960), pp. 21–24.

†Not reproduced here.

line is basically the same. The Contracting Parties have decided to determine concretely the boundary between the two countries in the following ways in accordance with three different cases:

1. Sections where the delineation of the boundary line between the two countries on the maps of the two sides is identical

In these sections the boundary line shall be fixed according to the identical delineation on the maps of the two sides. The joint committee will send out joint survey teams composed of an equal number of persons from each side to conduct survey on the spot and erect boundary markers.

After the boundary line in these sections is fixed in accordance with the provisions of the above paragraph, the territory north of the line will conclusively belong to China, while the territory south of the line will conclusively belong to Nepal, and neither Contracting Party will any longer lay claim to certain areas within the territory of the other Party.

2. Sections where the delineation of the boundary line between the two countries on the maps of the two sides is not identical, whereas the state of actual jurisdiction by each side is undisputed

The joint committee will send out joint survey teams composed of an equal number of persons from each side to conduct survey on the spot, determine the boundary line and erect boundary markers in these sections in accordance with concrete terrain features (watersheds, valleys, passes, etc.) and the actual jurisdiction by each side.

3. Sections where the delineation of the boundary line between the two countries on the maps of the two sides is not identical and the two sides differ in their understanding of the state of actual jurisdiction

The joint committee will send out joint teams composed of an equal number of persons from each side to ascertain on the spot the state of actual jurisdiction in these sections, make adjustments in accordance with the principles of equality, mutual benefit, friendship and mutual accommodation, determine the boundary line and erect boundary markers in these sections.

ARTICLE IV

The Contracting Parties have decided that, in order to ensure tranquility and friendliness on the border, each side will no longer dispatch armed personnel to patrol the area on its side within twenty kilometres of the border, but only maintain its administrative personnel and civil police there.

ARTICLE V

The present Agreement is subject to ratification and the instruments of ratification shall be exchanged in Kathmandu as soon as possible.

The present Agreement will come into force immediately on the exchange of the instruments of ratification‡ and will automatically cease to be in force when the Sino-Nepalese boundary treaty to be signed by the two Governments comes into force.

Done in duplicate in Peking on the twenty-first day of March, 1960, in the Chinese, Nepalese and English languages, all texts being equally authentic.

(Signed) Chou En-lai Plenipotentiary of the Government of the People's Republic of China	(Signed) B. P. Koirala Plenipotentiary of His Majesty's Government of Nepal

‡Came into force Apr. 28, 1960.

Appendix D: Treaty of Friendship and Mutual Non-Aggression between the Kingdom of Afghanistan and the People's Republic of China*

(August 26, 1960)

His Majesty the King of Afghanistan and the Chairman of the People's Republic of China,

Desiring to maintain and further develop lasting peace and profound friendship between the Kingdom of Afghanistan and the People's Republic of China,

Convinced that the strengthening of good-neighbourly relations and friendly cooperation between the Kingdom of Afghanistan and the People's Republic of China conforms to the fundamental interests of the peoples of the two countries and is in the interest of consolidating peace in Asia and the world,

Have decided for this purpose to conclude the present Treaty in accordance with the fundamental principles of the United Nations Charter and the spirit of the Bandung Conference, and have appointed as their respective plenipotentiaries:

His Majesty the King of Afghanistan:

Deputy Prime Minister and Minister of Foreign Affairs Sardar Mohammed Naim,

The Chairman of the People's Republic of China:

Vice-Premier of the State Council and Minister of Foreign Affairs Chen Yi.

The above-mentioned Plenipotentiaries, having examined each other's credentials and found them in good and due form, have agreed upon the following:

ARTICLE I

The Contracting Parties recognize and respect each other's independence, sovereignty and territorial integrity.

ARTICLE II

The Contracting Parties will maintain and develop peaceful and friendly relations between the two countries. They undertake to settle all disputes between them by means of peaceful negotiation without resorting to force.

ARTICLE III

Each Contracting Party undertakes not to commit aggression against the other and not to take part in any military alliance directed against it.

TYC, 9:12–14; English trans. in Chung-hua jen-min kung-ho kuo wai-chiao pu pien (ed.), *Chung-hua jen-min kung-ho kuo yu-hao t'iao-yüeh hui-pien* (Collection of friendship treaties of the People's Republic of China) (Peking, 1965), pp. 22–24.

ARTICLE IV

The Contracting Parties have agreed to develop and further strengthen the economic and cultural relations between the two countries in a spirit of friendship and cooperation and in accordance with the principles of equality and mutual benefit and of non-interference in each other's internal affairs.

ARTICLE V

The present Treaty is subject to ratification and the instruments of ratification will be exchanged in Peking as soon as possible.

The present Treaty will come into force immediately on the exchange of the instruments of ratification† and will remain in force for a period of ten years.

Unless either of the Contracting Parties gives to the other notice in writing to terminate it at least one year before the expiration of this period, it will remain in force indefinitely, subject to the right of either Party to terminate it after it has been valid for ten years by giving to the other in writing notice of its intention to do so one year before its termination.

Done in duplicate in Kabul on the twenty-sixth day of August 1960, in the Persian, Chinese and English languages, all texts being equally authentic.

(Signed) Mohammad Naim
Plenipotentiary of the Kingdom of Afghanistan

(Signed) Chen Yi
Plenipotentiary of the People's Republic of China

†Came into force Dec. 12, 1960.

Appendix E: Trade and Payments Agreement between the Government of Ceylon and the Government of the People's Republic of China*

(October 3, 1962)

The Government of Ceylon and the Government of the People's Republic of China, for the purpose of further developing the friendship between the Governments and the peoples of the two countries and of strengthening the economic and trade relations between the two countries have, on the basis of equality and mutual benefit, reached agreement as follows:—

ARTICLE I

The two Contracting Parties will take all appropriate measures to develop trade between their two countries and agree to facilitate the exchange of goods between the two countries.

ARTICLE II

The trade between the two countries shall be based on the principle of a balance between the values of imports and exports.

ARTICLE III

The two annexed Schedules A and B† which constitute an integral part of this Agreement show the export commodities of each country. This Agreement shall not preclude trade in commodities not mentioned in the annexed Schedules A and B.

ARTICLE IV

The two Contracting Parties shall, before the end of October each year, conclude a protocol of the commodities to be exchanged between the two parties in the following calendar year. This protocol shall specify:

(1) The aggregate value together with the names and approximate quantities of the commodities which the two Contracting Parties will undertake to import and export during the year covered by the protocol, and

(2) The aggregate value together with the names and approximate quantities of the commodities which the two Contracting Parties will endeavour to import and export during the year covered by the protocol.

ARTICLE V

The prices of commodities to be imported and exported under this Agreement shall be fixed at international market price levels.

*_TYC_, 11:115–118; English trans. in [Ceylon] Treaty Series, 7:15–17 (1965).
†Not reproduced here.

ARTICLE VI

The exchange of goods between the two countries shall be carried out in accordance with the import and export and foreign exchange regulations in force from time to time in each country.

ARTICLE VII

The two contracting Parties agree that trade under this Agreement, including trade under the protocols signed in terms of Article IV, may be conducted through the state trading organisations of Ceylon and China as well as through other importers and exporters in the two countries.

ARTICLE VIII

The two Contracting Parties will grant to each other most-favoured-nation treatment in respect of the issue of import and export licenses, and the levy of customs duties, taxes, and any other charges imposed on or in connection with the importation, exportation and transhipment of commodities, subject to the following exceptions:–

(1) Any special advantages which are accorded or may be accorded in the future by either of the Contracting Parties to contiguous countries in order to facilitate frontier trade and,

(2) Any special advantages which are accorded or may be accorded in the future under any preferential system of which either of the Contracting Parties is or may become a member.

ARTICLE IX

The two Contracting Parties agree that the payment arrangements between the two countries under this Agreement shall be in accordance with the following terms:

(1) The Government of Ceylon shall open two accounts in the People's Bank of China, Peking, styled Government of Ceylon Account "A" and Government of Ceylon Account "B".

The Government of China shall open two accounts in the Central Bank of Ceylon, Colombo, styled Government of China Account "A" and Government of China Account "B".

The above accounts shall bear no interest and shall be free of charges.

(2) Payments for the purchase of commodities which the two Contracting Parties have undertaken to import and export in terms of the yearly protocol referred to in Article IV of this Agreement, and payments for the relative incidental expenses, shall be made through the "A" accounts mentioned in paragraph (1) above.

Payments for other purchases and the relative incidental expenses as well as other payments approved by the Foreign Exchange Control authorities of both countries shall be made through the "B" accounts mentioned in paragraph (1) above. The phrase "relative incidental expenses" shall mean the expenses of services in connection with the exchange of goods such as transport charges including charter hire of ships and connected expenses, insurance, arbitration awards, warehousing and customs fees, agents' commissions, advertising, brokerage and other such charges.

(3) The accounts specified in paragraph (1) above shall be maintained in Ceylon rupees.

(4) Any residual balances in the "A" accounts specified in paragraph (1) above, outstanding on 31st March of the succeeding year, shall be settled by payment in Pound Sterling or any other currency mutually acceptable immediately after the accounts have been reconciled.

Payments in respect of contracts entered into under the annual protocol of any year, which are made after 31st March of the succeeding year shall be brought to account under the "A" account of the succeeding year.

(5) The Balances in the "B" accounts specified in paragraph (1) above shall be reviewed once every quarter by the two Contracting Parties for the purpose of ensuring that trade between the two countries progresses in balance.

Any balances in the "B" accounts remaining outstanding at the end of each calendar year, shall be settled as far as possible by delivery of goods during the first three months of the succeeding year. Any residual balances in the "B" accounts still remaining outstanding on 31st March of the succeeding year, shall be settled by payment in pound sterling or any other currency mutually acceptable immediately after the accounts have been reconciled.

(6) The exchange rate for settlement of balances contemplated in paragraphs (4) and (5) above shall be the middle of the Central Bank of Ceylon's buying and selling rates for Pound Sterling or other currency at the time of payment.

(7) The Central Bank of Ceylon and the People's Bank of China shall work out the technical details necessary for the implementation of this Article.

ARTICLE X

This Agreement shall come into force on 1st January, 1963 and shall remain in force for a period of Five Years. This Agreement may be extended by negotiation of both parties three months before its expiration.

This Agreement is signed in Peking this 3rd day of October, 1962, in two copies, each written in the English and Chinese Languages, and both texts being equally authentic.

(Sgd.) T. B. Ilangaratne,
for the Government of Ceylon.

Yeh Chih Chong,
for the Government of the People's Republic of China.

Appendix F: Protocol Relating to the Exchange of Commodities Between Ceylon and the People's Republic of China for 1966*

(October 12, 1965)

In accordance with the provisions of the "Trade and Payments Agreement between the Government of Ceylon and the Government of the People's Republic of China" signed on the 3rd day of October 1962, representatives of the Government of Ceylon and the Government of the People's Republic of China after negotiations in Peking on the exchange of commodities between Ceylon and China for 1966 have reached agreement as follows:

ARTICLE I

The Government of the People's Republic of China undertakes to buy and the Government of Ceylon undertakes to sell Ceylonese commodities set out in Schedule "A 1"† of this Protocol; and the Government of Ceylon undertakes to buy and the Government of the People's Republic of China undertakes to sell Chinese commodities set out in Schedule "A 2" of this Protocol.

ARTICLE II

The Government of Ceylon and the Government of the People's Republic of China will endeavour to expand trade between their two countries on the basis of maintaining a balance between the values of imports and exports, and do their utmost to increase their respective imports and exports to the extent of the values set out in Schedules "B 1" and "B 2" attached to this Protocol. These amounts will include the values of the commodities contemplated by Article I.

ARTICLE III

In order to facilitate the implementation of this Protocol, the state trading organisations of Ceylon and China as well as other importers and exporters in the countries may conclude contracts for the commodities listed in Schedules "A 1", "A 2", "B 1" and "B 2" of this Protocol.

ARTICLE IV

This Protocol shall come into force on the 1st January, 1966, and remain valid for a period of one year.

It is mutually agreed that the contracts concluded in accordance with Article III of the Protocol may continue to be in force until the expiry of their respective terms of validity even after the termination of this Protocol.

*[Ceylon] Treaty Series, 11:8 (1965).

†"A 1," "A 2," "B 1," and "B 2" are not reproduced here.

This Protocol is signed in Peking this 12th day of October 1965, in three copies, each written in the Sinhala, Chinese and English languages, all three texts being equally authentic.

For the Government Ceylon,
(Sgd.) M. V. P. Peiris.

For the Government of the People's Republic of China,
(Sgd.) Lin Hai Yun.

Appendix G: Exchange of Notes between the People's Republic of China and the Republic of Finland on Trade-marks Registration*

(January 26, 1967)

1. Note from the Ministry of Foreign Affairs of the People's Republic of China to the Embassy of the Republic of Finland in China.

   Peking, January 26, 1967

   The Ministry of Foreign Affairs of the People's Republic of China presents its compliments to the Embassy of the Republic of Finland in China, and has the honour to state the following:

   Concerning the question of mutual reciprocal registration of trade-marks between the People's Republic of China and the Republic of Finland, the Government of the People's Republic of China considers that, provided citizens, companies and co-operatives of the People's Republic of China can obtain registration of trade-marks in the Republic of Finland in accordance with Finnish laws and there be granted exclusive use of such trade-marks, citizens, companies and co-operatives of the Republic of Finland can obtain registration of trade-marks in the People's Republic of China in accordance with Chinese laws and there be granted exclusive use of such trade-marks. However, the trade-marks of one country may not be protected in the other country to a larger extent or during a longer period than in the first mentioned country.

   If the Government of the Republic of Finland shares the opinion of the Government of the People's Republic of China expressed above and gives a corresponding reply to the Ministry of Foreign Affairs, each party shall immediately carry out procedures concerning application for registration of trade-marks by citizens, companies and co-operatives of the other party in accordance with the laws valid in its country.

   The Ministry of Foreign Affairs avails itself of this opportunity etc.

2. Note from the Embassy of the Republic of Finland in China to the Ministry of Foreign Affairs of the People's Republic of China.

   Peking, January 26, 1967

   The Finnish Embassy present their compliments to the Ministry of Foreign Affairs of the People's Republic of China and acknowledge receipt of the note of the Ministry dated January 26th, 1967, and reading as follows: [text of note 1].

   In reply, the Finnish Embassy have the honour to inform the Ministry of Foreign Affairs of the People's Republic of China that the Government of the Republic of Finland share the opinion of the Government of the People's Republic of China expressed in the

* *Suomen Asetuskokoelman Sopimussarja: Ulkovaltain Kanssa Tehdyt Sopimukset, 1967* (Helsinki, 1968), XXII, 22–23.

note cited above. Each part shall thus, under the aforesaid premises and according to the laws valid in its country, immediately effect registration of trade-marks for the account of citizens, companies and co-operatives of the other party.

The Finnish Embassy avail themselves of this opportunity [etc.]

Appendix H: Joint Communiqué of the People's Republic of China and the Republic of Zambia* (Excerpt)

(June 25, 1967)

At the invitation of Chou En-lai, Premier of the State Council of the People's Republic of China, His Excellency Kenneth David Kaunda, President of the Republic of Zambia, and Madame Kaunda paid a state visit to the People's Republic of China from June 21 to 25, 1967.

Accompanying them on the visit were: Hon. E.H.K. Mudenda, Minister of Agriculture; Hon. H.D. Banda, Minister of Co-operatives, Youth and Social Development; Hon. J.H. Monga, Minister of State for the Barotse Province; Mr. A. Chalikulima, Assistant to the Minister of State for the Western Province and senior officials of the Government of Zambia.

The Minister of Finance of Zambia, the Hon. A.N.L. Wina was at the same time on a visit to Peking at the head of a delegation to negotiate an economic and technical co-operation agreement.

Chairman Mao Tse-tung, the great leader of the Chinese people, met His Excellency President Kenneth David Kaunda and Madame Kaunda and the other Zambian friends during their visit in China. Vice-Chairman Lin Piao, the close comrade-in-arms of Chairman Mao Tse-tung, was present at the meeting. Chairman Mao Tse-tung and Vice-Chairman Lin Piao had a cordial and friendly conversation with His Excellency the President.

His Excellency President Kenneth David Kaunda and Madame Kaunda and the other Zambian friends visited a people's commune, factories, a university and a unit of the Chinese People's Liberation Army in Peking and Shanghai, and had extensive friendly contacts with the masses of workers, peasants and soldiers, revolutionary teachers and students and Red Guards of China engaged victoriously in the great proletarian cultural revolution unparalleled in history. They were accorded a warm welcome and cordial reception by the Chinese Government and people. This fully demonstrated the profound friendship between the Chinese and Zambian peoples.

Premier Chou En-lai held talks with His Excellency President Kenneth David Kaunda.

Present at the talks on the Chinese side were: Chen Yi, Vice-Premier of the State Council and Minister of Foreign Affairs; Li Hsien-nien, Vice-Premier of the State Council and Minister of Finance; Lin Hai-yun, Acting Minister for Foreign Trade; Hsu Yi-hsin, Vice-Minister of Foreign Affairs; Hsieh Huai-teh, Vice-Chairman of the Commission for Economic Relations with Foreign Countries; Chu Tu-nan, Vice-Chairman of the Commission for Cultural Relations with Foreign Countries; Chin Li-chen, Chinese Ambassador to Zambia; and others.

Present at the talks on the Zambian side were: Hon. A.N.L. Wina, Minister of Finance; Hon. E.H.K. Mudenda, Minister of Agriculture; Hon. H.D. Banda, Minister of Co-operatives, Youth and Social Development;

*PR, 10:12–14, no. 27 (June 30, 1967).

Hon. J. H. Monga, Minister of State for the Barotse Province; Mr. A. Chalikulima, Assistant to the Minister of State for the Western Province; and senior officials of the Government of Zambia.

The talks proceeded in a cordial and friendly atmosphere. During the talks the two sides had a full exchange of views on international issues of common interest and on the further development of friendly relations and co-operation between the two countries, and satisfactory results were achieved.

The two sides noted with great satisfaction that there has been a rapid development in the friendly relations and co-operation between China and Zambia in the political, economic, trade and cultural fields since the two countries established diplomatic relations. During His Excellency President Kenneth David Kaunda's current visit, China and Zambia have signed an agreement on economic and technical co-operation, which will further promote the development of the friendly relations and co-operation between the two countries. The two sides have agreed to exert every effort to consolidate and develop the friendly relations between China and Zambia and strengthen the friendship between the two peoples.

The Chinese side reaffirms its full respect for the policy of non-alignment pursued by the Government of Zambia in international affairs, and its firm support for the struggle waged by the Republic of Zambia under the leadership of His Excellency President Kaunda to safeguard national independence and state sovereignty, to develop national economy and culture and to oppose imperialism, colonialism, neo-colonialism and racial discrimination. The Chinese side warmly commends the contributions made by the Zambian Government and people in supporting the African national-liberation movement and promoting Afro-Asian solidarity.

The two sides strongly oppose the British connivance with the Rhodesian and other colonial authorities in perpetuating white domination south of the Zambezi.

Chinese side firmly supports the Government and people of Zambia in their just struggle against the British Government's Rhodesian policy and the Rhodesian colonial authorities.

The Zambian side extends warm congratulations upon the great victories of the great proletarian cultural revolution initiated and led personally by Chairman Mao Tse-tung, the great leader of the Chinese people. It highly appraises the great achievements won by the Chinese people in the cause of socialist construction. It reaffirms its support for the restoration of the legitimate rights of the People's Republic of China in the United Nations, its opposition to all schemes of creating "two Chinas," and its support for the Chinese people in their just struggle to safeguard national sovereignty and territorial integrity. The Chinese side expresses its thanks for all this.

The Zambian side warmly congratulates China on the successful explosion of her first hydrogen bomb and regards this as a tremendous contribution to the safeguarding of world peace. The two sides express their willingness to work together with all the other peace-loving people and countries of the world for the noble aim of completely prohibiting and thoroughly destroying nuclear weapons.

Both sides maintain that the destiny of mankind should never be controlled by one or two powers which have a monopoly of nuclear weapons, but should be decided by all countries in the world . . .

The two sides express great concern over the Vietnam question. The Chinese side resolutely supports the heroic Vietnamese people in their struggle against U.S. aggression and for national salvation, strongly condemns U.S. imperialism for its frenzied act of expanding the war of aggression in Vietnam, firmly demands that the troops of the United States and its vassals must withdraw from south Vietnam completely, and holds that the South Vietnam National Front for Liberation must be recognized as the sole genuine representative of the south Vietnamese people. The two sides are of the opinion that the Vietnamese people's basic national rights, namely, independence, sovereignty, unification and territorial integrity, must be respected and that the Vietnam question should be settled by the Vietnamese people themselves.

The two sides firmly support the people of Mozambique, Angola, Guinea (Bissau), Southwest Africa, Swaziland, French Somaliland and other African countries still under colonial rule in their struggle for national independence . . .

The two sides note with satisfaction that His Excellency President Kenneth David Kaunda's current visit to China has enhanced the friendship and mutual understanding between the people of the two countries, ushered in a new stage of development in the relations of friendship and co-operation between the two countries and made important contributions to the strengthening of Afro-Asian solidarity.

His Excellency President Kenneth David Kaunda of the Republic of Zambia extended an invitation to Premier Chou En-lai of the State Council of the People's Republic of China for a visit to the Republic of Zambia at a time convenient to him. Premier Chou En-lai accepted the invitation with pleasure.

June 25, 1967

Appendix I: Agreement between Fishery Association of China and Japan-China Fishery Association concerning Fisheries in Yellow Sea and East China Sea*

(December 17, 1965)

According to the principles of equality, mutual benefit and friendly cooperation and for the sake of making rational use of the fishing grounds in the Yellow Sea and the East China Sea, preserving fishery resources and avoiding disputes between fishing vessels of both sides operating there, so as to improve the friendly relations between the fishery circles of China and Japan, the delegations sent respectively by the Fishery Association of China and the Japan-China Fishery Association (hereinafter referred to as the fishery associations of both sides for short) have, after due negotiations, reached agreement as follows:

ARTICLE 1

The area of sea applicable to this agreement (hereinafter called the agreed sea area for short) is the area of high seas in the Yellow Sea and the East China Sea east of the line joining in proper sequence the points – Lat. 39° 45′ N., Long. 124° 9′ 12″ E.; Lat. 37° 20′ N., Long. 123° 3′ E.; Lat. 36° 48′ 10″ N., Long. 122° 4′ 30″ E.; Lat. 35° 11′ N., Long. 120° 38′ E.; Lat. 30° 44′ N., Long. 123° 25′ E.; Lat. 29° N., Long. 122° 45′ E.; Lat. 27° 30′ N., Long. 121° 30′ E.; Lat. 27° N., Long. 121° 10′ E. – and north of Lat. 27° N.

ARTICLE 2

(1) For the six fishing areas within the agreed sea area, the fishery associations of both sides shall separately stipulate the maximum number of trawlers (including double-dragger and single-boat purse-seines, hereinafter called fishing vessels for short) of the Chinese and Japanese sides (hereinafter called the two sides for short) actually engaged in fishing at a specific period. For particulars, see Appendix 1.†

(2) The provisions in this article shall place no restriction on navigation in the agreed sea area.

ARTICLE 3

For the preservation of fishery resources, simultaneously with restricting the mesh of trawl-nets used by fishing vessels of both sides, the fishery associations of both sides shall also place restriction on the haul of young fish among the important species for fishing which are considered necessary to be placed under special protection – according to the provisions of Appendix 2.

*_People's Daily_, Dec. 18, 1965; trans. in _SCMP_, 3613:27–29 (Jan. 10, 1966).

†The appendixes are not reproduced here.

ARTICLE 4

In order to insure safety production on the sea between trawlers and between trawlers and other types of fishing vessels and to preserve normal order, the trawlers of both sides shall abide by the provisions contained in Appendix 3.

ARTICLE 5

(1) When the fishing vessels of one side encounter marine disaster or other irresistible calamities, or when the crew aboard them are seriously wounded or critically ill, and it is necessary for them to seek emergency shelter or assistance, the fishery association of the other side and its fishing vessels in the fishing ground should do everything possible for their assistance and rescue.

(2) When the fishing vessels of one side enter and anchor in a port of the other side in an emergency, they should abide by the provisions of Appendix 4.

ARTICLE 6

For the sake of preserving fishery resources and developing fishery production of the two sides, the fishery associations of both sides are willing to exchange data concerning fishery investigation and research and technical improvement, and scientific and technical personnel in fisheries. For particulars, see Appendix 5.

ARTICLE 7

(1) When the vessels of one side discover that the provisions of Article 2 have been violated by the fishing vessels of the other side, they should notify the fishery association of the other side through their own fishery association. The fishery association of the side thus notified should take prompt, effective and proper action to deal with the complaints, and notify the fishery association of the other side of the results.

(2) When disputes between the two sides arise due to the collision of trawlers or trawlers with other types of fishing vessels, or to the damage of fishing gears, steps should promptly be taken to safeguard the safety of the damaged vessels, and whenever possible, consultations should be conducted on the spot, and written statements on the circumstances of the incidents exchanged. After that, reports should be made to their respective fishery associations, and the fishery associations of the two sides shall look into the actual circumstances and negotiate for settlement.

ARTICLE 8

The appendixes to this agreement and text of the agreement shall be equally binding.

ARTICLE 9

The fishery associations of both sides shall be responsible for the implementation of this agreement.

ARTICLE 10

This agreement shall come into force on December 23, 1965 and shall be good for two years.

This agreement is signed on December 17, 1965 at Peking in two copies. Both the Chinese language and the Japanese language are used in each copy, and both languages shall have the same validity.

The Delegation of the Fishery Association of China

Head: Li Ying-shu
Deputy Head: Wang Yün-hsiang
Adviser: Chao An-po
Members: Keng Ju-yün, Han Chü-li
Ch'ang Fei-hu, Wang Yeh-yü
Shao Tsai-fu, Ch'en Hsin

The Delegation of the Japan-China Fishery Association

Head: Tokushima Kitaro
Adviser: Eguchi Jiku
Members: Kawakami, Kodera
Moriguchi, Ikiri Kihira
Takihisashi Saburo, Kawahon
Nakamura Masamichi

Selected Bibliography

"Afro-Asian Writers' Bureau Denounces 'Treaty on Non-Proliferation of Nuclear Weapons,'" *PR*, 11:6, no. 29 (July 19, 1968).

"Albanian People's Assembly Passes Law Denouncing Warsaw Treaty, Announces Albania's Withdrawal from It," *PR*, 11:7, no. 38 (Sept. 20, 1968).

American Foreign Policy, 1950–1955, Basic Documents, vol. I, Department of State pub. 6446. Washington, D.C.: U.S. Government Printing Office, 1957.

American Journal of International Law. Washington, D.C.: American Society of International Law, 1907 —.

"At Tirana Mass Rally: Speech by Comrade Huang Yung-sheng (Excerpt)," *PR*, 11:7–9, no. 50 (Dec. 13, 1968).

Avakov, Mirza Mosesovich. *Pravopreemstvo sovetskogo gosudarstva* (Legal succession of the Soviet State). Moscow: Gos. izd-vo iurid. lit-ry, 1961.

"Behind the Hue and Cry over the 'Hungarian Issue'" (editorial), *People's Daily*, Nov. 14, 1956, p. 1.

Blix, H. *Treaty-Making Power*. New York: Praeger, 1959.

Briggs, H. W., ed. *The Law of Nations: Cases, Documents, and Notes*, 2nd ed. New York: Appleton-Century-Crofts, 1952.

British and Foreign State Papers. Vol. 78 (1886–1887), London: William Ridgway, 1894. Vol. 100 (1906–1907), London: His Majesty's Stationery Office, 1911. Vol. 161 (1954), London: Her Majesty's Stationery Office, 1963. Vol. 164 (1959–1960), London: Her Majesty's Stationery Office, 1967.

Brownlie, I. *Principles of Public International Law*. London: Oxford University Press, 1966.

Buell, Raymond L. "The Termination of Unequal Treaties," *Proceedings of the American Society of International Law: Twenty-First Annual Meeting* (1927), pp. 90–99.

Calvo, Carlos. *Le Droit international théorique et pratique*, vol. I, 3rd ed. Paris: A. Rousseau, 1880.

Calvocoressi, Peter. *Survey of International Affairs 1953*. London and New York: Oxford University Press, 1956.

Carlyle, Margaret, ed. *Documents on International Affairs 1947–1948*. New York and London: Oxford University Press, 1952.

"Ceremonies in Peking – Exchange Students Expelled," *CDSP*, 18:14, no. 40 (Oct. 26, 1966).

"Chairman Mao Tse-tung Tells the Delegation of the Japanese Socialist Party that the Kuriles Must Be Returned to Japan," *Sekai Shūhō* (Tokyo), Aug. 11, 1964.

Chao Yüeh. "A Preliminary Criticism of Bourgeois International Law," *KCWTYC*, 3:1–9 (1959).

Chen, Charng-ven. "Communist China's Attitude Toward Consular Relations," unpub. paper. Harvard Law School, 1970.

Ch'en T'i-ch'iang. "The Hungarian Incident and the Principle of Non-Intervention," *Enlightenment Daily*, Apr. 5, 1957, p. 1.

——"The Illegality of Atomic Weapons from the Viewpoint of International Law," *Shih-chieh chih-shih* (World knowledge), 4:11–12 (Feb. 20, 1955).

——"Sovereignty of Taiwan Belongs to China," *People's Daily*, Feb. 8, 1955, p. 4.

——"Taiwan – A Chinese Territory," *Law in the Service of Peace* (Brussels), 5:38–44 (1956).

——"Unconditional Repatriation – An Inviolable Principle of the Geneva Convention," *PC*, 2:26–28 (Jan. 16, 1953).

Cheng-fa yen-chiu (Studies in political science and law). Peking: Chung-kuo cheng-chih fa-lu hsüeh-hui, 1954–1966.

Chiang Ching-Kuo. *Fu-chung chih-yüan* (Carrying heavy burdens to a great distance). Taipei: Yu-shih shu-tien, 1963.

Chiang Kai-shek. *China's Destiny*, trans. Wang Chung-hui from 1944 Chinese ed. New York: Macmillan, 1947.

——*Soviet Russia in China*, rev. ed. New York: Farrar, Straus and Cudahy, 1958.

Chiao-hsüeh yü yen-chiu (Teaching and research). Peking: Chung-kuo jen-min ta-hsüeh, 1957–1959.

Ch'ien Szu. "A Criticism of the Views of Bourgeois International Law on the Question of the Population," *KCWTYC*, 5:40–49 (1960).

Ch'ien T'ai. *Chung-kuo pu-p'ing teng t'iao-yüeh chih yüan-ch'i chi fei-ch'u chih ching-kuo* (The origin and abolition of China's unequal treaties). Taipei: Kuo-fan yen-chiu yüan, 1961.

Chin Szu-k'ai. *Communist China's Relations with the Soviet Union, 1949–1957*. Hong Kong: Union Research Institute, 1961.

Ch'in Tzu-ch'iang. *Lun Mei-Chiang ch'in-lüeh t'iao-yüeh* (On the American-Chiang Kai-shek treaty of aggression). Peking: Shih-chieh chih-shih she, 1955.

"China Announces It Has Reached Accord with USSR to Hold Negotiations on Border Dispute," *New York Times*, Oct. 8, 1969, p. 1.

"China-Mali Friendship Treaty," *PR*, 8:25, no. 18 (Apr. 30, 1965).

"China and Pakistan Sign Economic and Technical Co-operation Agreement," *PR*, 12:26, no. 2 (Jan. 10, 1969).

"China Protests at Hungary's Not Allowing Chinese Students to Continue Their Studies," *SCMP*, 3826:30–31 (Nov. 23, 1966).

"China Protests at U.S. Action in South Korea," *SCMP*, 1562:29–30 (July 3, 1957).

"China Resumes Supply of Water to Hong Kong," *New York Times*, Oct. 1, 1967, p. 6.

"China Strongly Protests against Worsening of Sino-Ghanaian Relations by Ghanaian Authorities," *PR*, 9:8–10, no. 13 (March 25, 1966).

"China, U.S.S.R. Discuss 1958 Cultural Cooperation Plan," *SCMP*, 1659: 39–40 (Nov. 26, 1957).

"China's State Council Approves Sino-Mauritanian Agreements," *SCMP*, 3907:38 (Mar. 29, 1967).

"Chinese Embassy Protests against Burmese Government's Sabotage of Economic-Technical Cooperation Agreement," *SCMP*, 3977:27–28 (July 11, 1967).

"Chinese Foreign Ministry Lodges Strongest Protest Against Soviet Authorities for Unjustifiably Suspending Studies of All Chinese Students in USSR," *SCMP*, 3809:38–40 (Oct. 27, 1966).

"Chinese Foreign Ministry Refutes Mongolian Government's Anti-China Statement," *PR*, 10:25–26, no. 35 (Aug. 25, 1967).

"Chinese Foreign Ministry Sends Protest Note to Bulgarian Embassy," *SCMP*, 3959:38–39 (June 14, 1967).

"Chinese Government Demands Withdrawal of U.S. Forces from Lebanon," *PR*, 1:7, no. 22 (July 22, 1958).

"Chinese Government Statement: China Will Never Recognize 'ROK-Japan Basic Treaty,' June 26," *PR*, 8:7–8, no. 27 (July 2, 1965).

"Chinese Government Statement: China's Aid to Vietnam in Fighting U.S. Aggression Further Ceases to Be Subject to Any Bounds or Restrictions," *PR*, 9:19–20, no. 28 (July 8, 1966).

"Chinese Government Statement: Most Strongly Protesting against the Ne Win Government's Expulsion of Chinese Experts," *PR*, 10:27–28, no. 46 (Nov. 10, 1967).

"Chinese Government Supports Soviet Proposal on Berlin Question," *SCMP*, 1921:45–47 (Dec. 24, 1958).

Chinese Ministry of Information. *The Collected Wartime Messages of Generalissimo Chiang Kai-shek 1937–1945*, 2 vols. New York: John Day, 1946.

Chinese People's Institute of Foreign Affairs. *New Development in Friendly Relations between China and Nepal.* Peking: Foreign Languages Press, 1960.

——*Oppose U.S. Occupation of Taiwan and "Two Chinas" Plot: A Selection of Important Documents.* Peking: Foreign Languages Press, 1958.

——*A Victory for the Five Principles of Peaceful Coexistence.* Peking: Foreign Languages Press, 1960.

"Chinese Protest to U.S. Company on Sinking of Fishing Vessel," *SCMP*, 2782:33 (July 20, 1962).

Chiu, Hungdah. "The Attitude of Chinese Communists Toward the 'Question of Taiwan,'" *Jen-wu* (Character magazine [Hong Kong]), 9:6–8, 38–39 (December 1967).

——*The Capacity of International Organizations to Conclude Treaties and the Special Legal Aspects of the Treaties So Concluded.* The Hague: Martinus Nijhoff, 1966.

——"Certain Legal Aspects of Communist China's Treaty Practice," *Proceedings of the American Society of International Law: Sixty-First Annual Meeting* (1967), pp. 117–126.

——"Chung-hua jen-min kung-ho kuo t'iao-yüeh chi" (Compilation of treaties of the People's Republic of China, vols. I–XIII), book review, *AJIL*, 61:1095–1098 (1967).

——"Communist China and the Law of Outer Space," *International and Comparative Law Quarterly*, 16:1135–1138 (1967).

——"Communist China's Attitude Toward International Law," *AJIL*, 60:245–267 (1966).

——"A Comparative Study of the Chinese and Western Positions on the Problem of Unequal Treaties," *Cheng-ta fa-hsüeh p'ing-lun* (Chengchi law review [Taipei]), 1:1–9 (December 1969).

——"Comparison of the Nationalist and Communist Chinese Positions on the Problem of Unequal Treaties," in J. A. Cohen, ed., *China's Practice of International Law: Some Case Studies* (to be published in 1971).

——"The Development of Chinese International Law Terms and the Problem of Their Translation into English," *Journal of Asian Studies*, 27:485–501 (1968).

——"Peiping's Role in Communist Inter-Governmental Organizations," *Issues and Studies* (Taipei), 3:33–35, no. 9 (June 1967).

——"Suspension and Termination of Treaties in Communist China's Theory and Practice," *Osteuropa-Recht* (West Germany), 15:169–190 (1969).

——"The Theory and Practice of Communist China with Respect to the Conclusion of Treaties," *Columbia Journal of Transnational Law*, 5:1–13 (1966).

Chiu, Hungdah, assisted by R. Randle Edwards. "Communist China's Attitude toward the United Nations: A Legal Analysis," *AJIL*, 62:20–50 (1968).

Ch'iu Jih-ch'ing. "Further Discussion of the System of International Law at the Present Stage," *FH*, 3:40–43 (1958).

Chou Fu-lun. "On the Nature of Modern International Law," *CHYYC*, 3:52–56 (1958).

Chou Keng-sheng. "China's Legitimate Rights in the United Nations Must Be Restored," *People's Daily*, Dec. 5, 1961, p. 5.

——"Don't Allow American and British Aggressor to Intervene in the Internal Affairs of Other States," *CFYC*, 4:3–4 (1958).

——"Looking at the West Berlin Question from the Angle of International Law," *KCWTYC*, 1:40–45 (1959).

——"The Persecution of Chinese Personnel by Brazilian Coup d'Etat Authority Is a Serious International Illegal Act," *People's Daily*, Apr. 24, 1964, p. 4.

——"The Principles of Peaceful Coexistence from the Viewpoint of International Law," *CFYC*, 6:37–41 (1955).

Chou Tze-ya. "Talks on the Question of Suez Canal," *Hua-tung cheng-fa hsüeh-pao* (East China journal of political science and law), 3:11, 35–38 (1956).

Chu Li-sun. "The Use of Atomic and Hydrogen Weapons Is the Most Serious Criminal Act in Violation of International Law," *CFYC*, 4:30–33 (1955).

Chung Ho. "How U.S. Imperialism Has Torn Up the Geneva Agreements," *PR*, 8:24–25, no. 14 (Apr. 2, 1965).

Chung-hua jen-min kung-ho kuo tui-wai kuan-hsi wen-chien chi (Compilation of documents relating to the foreign relations of the People's Republic of China), vols. I–X. Peking: Shih-chieh chih-shih ch'u-pan she, 1957–1965.

Chung-hua jen-min kung-ho kuo wai-chiao pu pien. *Chung-hua jen-min kung-ho kuo t'iao-yüeh chi* (Compilation of treaties of the People's Republic of China), vols. 1–13. Vols. 1–10, Peking: Fa-lü ch'u-pan she, 1957–1962; Vols. 11–13, Peking: Shih-chieh chih-shih ch'u-pan she, 1963–1965.

——*Chung-hua jen-min kung-ho kuo yu-hao t'iao yüeh hui-pien* (Collection of friendship treaties of the People's Republic of China). Peking: Shih-chieh chih-shih ch'u-pan she, 1965.

——*I-chiu-szu-chiu nien pa-yüeh shih-erh jih jih-nei-wa kung-yüeh* (The Geneva Conventions of 12 August 1949). Peking: Fa-lü ch'u-pan she, 1958.

Chung-hua min-kuo k'ai-kuo wu-shih nien wen-hsien (Documents and materials relating to the fiftieth anniversary of the Republic of China), 1st ser., vol. VII, pt. 1: *Ch'ing-ting chih kai-ke yü fan-tung* (The reform and anti-reform measures of the Ch'ing government). Taipei: Chung-hua min-kuo k'ai-kuo wu-shih nien wen-hsien pien-ch'uan wei-yuan hui, 1965.

——, appendixes: *Kung-fei huo-kuo shih-liao hui-pien* (Collection of historical materials relating to the Communist bandits' rebellion), vol. I. Taipei: Chung-hua min-kuo k'ai-kuo wu-shih nien wen-hsien pien-ch'uan wei-yuan hui, 1964.

Chung-shan ch'üan-shu (Complete works of Sun Yat-sen), vol. II. Shanghai: Ta-hua shu-chü, 1927.

Chung-yang jen-min cheng-fu fa-chih wei-yuan-hui pien. *Chung-yang jen-min cheng-fu fa-ling hui-pien* (Collection of laws and decrees of the Central People's Government), 7 vols., vol. I (1949–1950). Peking: Hsin-hua shu-tien, 1952.

Cohen, J. A. "Chinese Attitudes Toward International Law – And Our Own," *Proceedings of the American Society of International Law: Sixty-First Annual Meeting* (1967), pp. 108–116.

——"New Developments in Western Studies of Chinese Law: A Symposium," *Journal of Asian Studies*, 27:475–483 (1968).

Cohen, J. A., ed., *China's Practice of International Law: Some Case Studies*. To be published in 1971.

——*Contemporary Chinese Law: Research Problems and Perspectives*. Cambridge, Mass.: Harvard University Press, 1970.

"A Comment on the Statement of the Communist Party of the U.S.A.," (editorial), *People's Daily*, Mar. 8, 1963, p. 1.

Commentator. "Another Crime of the U.S.-Chiang Kai-shek Conspiracy," *People's Daily*, Feb. 19, 1966, p. 1.

——"Insidious U.S.-Soviet Collaboration Policy," *People's Daily*, July 23, 1967, p. 4.

——"A Nuclear Fraud Jointly Hatched by the United States and the Soviet Union," *People's Daily*, June 13, 1968, p. 5.

——"The Soviet Revisionist Renegades Try to Pull the Wool over the Eyes of the Public," *People's Daily*, Aug. 5, 1967, p. 6.

——"Vientiane Authorities Cannot Represent Laotian Peoples," *People's Daily*, June 23, 1965, p. 5.

"Communiqué on Sino-Indonesian Talk on Dual Nationality," *SCMP*, 1033:27 (Apr. 23–25, 1955).

"Communists Take U.S. Property in China," *DSB*, 22:119–122, no. 551 (Jan. 23, 1950).

"Conference of President Roosevelt, Generalissimo Chiang Kai-shek, and Prime Minister Churchill in North Africa," *DSB*, 9:393, no. 232 (Dec. 4, 1943).

"Continued Detention of U.S. Civilians by Communist China," *DSB*, 33:1049–1050, no. 861 (Dec. 26, 1955).

"Courageous and Resolute Revolutionary Action" (editorial), *People's Daily*, Sept. 20, 1968, p. 1.

"Crisis in Hong Kong Is Eased by Accord," *New York Times*, Dec. 1, 1967, pp. 1, 16.

Curl, Peter V., ed. *Documents on American Foreign Relations, 1954*. New York: Harper, 1955.

Current Digest of the Soviet Press. New York: Joint Committee on Slavic Studies, 1950—.

"Declaration of the Nationalist Government, June 16, 1928," *Chinese Social and Political Science Review*, 12:47–48, doc. sec. (1928).

Department of State Bulletin. Washington, D.C.: Government Printing Office, 1939—.

"Diabolical Social-Imperialist Face of the Soviet Revisionist Renegade Clique," *PR*, 11:8, no. 43 (Oct. 25, 1958).

"Document of the Ministry of Foreign Affairs of the People's Republic of China – Refutation of the Soviet Government's Statement of June 13, 1969, October 8, 1969," *PR*, 12:8–15, no. 41 (Oct. 10, 1969).

Documents of the United Nations Conference on International Organization: San Francisco, 1945, vol. XIII. New York and London: United Nations Information Organization in cooperation with Library of Congress, 1945.

Doolin, Dennis J. *Territorial Claims in the Sino-Soviet Conflict: Documents and Analysis*. Stanford: Hoover Institution on War, Revolution, and Peace, Stanford University, 1965.

Durdenevskii, V. N., and S. B. Krylov. "Narushenie imperialisticheskimi gosudarstvami printsipa sobliudeniia mezhdunarodnykh dogovorov," *Mezhdunarodnaia zhizn'* (Moscow), 2:64–71 (1955).

Edwards, R. Randle. "The Attitude of the People's Republic of China Toward International Law and the United Nations," *Papers on China*, 17:235–271 (1963).

"Facts on Sino-Cuban Trade," *PR*, 9:21–23, no. 3 (Jan. 14, 1966).

Fa-hsüeh (Science of law). Shanghai: Shang-hai fa-hsüeh hui [and] Hua-tung cheng-fa hsüeh-yuan, 1957–1959.

"Failure of Chinese Communists to Release Imprisoned Americans," *DSB*, 36:261–263, no. 921 (Feb. 18, 1957).

Fenwick, C. G. *International Law*, 4th ed. New York: Appleton-Century-Croft, 1965.

Fishel, Wesley R. *The End of Extraterritoriality in China*. Berkeley and Los Angeles: University of California Press, 1952.

Focsaneanu, L. "Les Grands Traités de la République populaire de Chine," *Annuaire francais de droit international, 1962*, pp. 139–177. Paris: Centre National de la Recherche Scientifique, 1963.

Folliot, Denise, ed. *Documents on International Affairs, 1952; 1954*. London and New York: Oxford University Press, 1955; 1957.

"Foreign Minister Chou En-lai's Statement on U.S.-British Draft Peace Treaty with Japan and San Francisco Conference," *NCNA*, Daily News Release no. 777 (Aug. 16, 1951), pp. 75–78.

Foreign Relations of the United States, Diplomatic Papers: The Conference of Berlin (The Potsdam Conference), 1945. Washington, D.C.: U.S. Government Printing Office, 1960.

Foreign Relations of the United States, Diplomatic Papers: The Conferences at Cairo and Teheran 1943. Washington, D.C.: U.S. Government Printing Office, 1961.

"The 4th Civil Trade Agreement between Japan and the People's Republic of China Signed at Peking, March 5, 1958," *Contemporary Japan*, 25:520–527 (1957–1958).

Fu Chu. "American Imperialism's Use of Poisonous Gas in South Vietnam Is a War Crime in Flagrant Violation of International Law," *People's Daily*, Apr. 3, 1965, p. 3.

"GAC Holds 102nd Meeting; Chang Han-fu Reports on Foreign Affairs," *SCMP*, 175:12 (Sept. 15–16, 1951).

Gittings, John. *Survey of the Sino-Soviet Dispute*. London: Oxford University Press, 1968.

Goodrich, L. M. "Korea: Collective Measures against Aggression," *International Conciliation*, 494:131–192 (October 1953).

Goodrich, L. M., and E. Hambro. *Charter of the United Nations: Commentary and Documents*, 2nd ed. London: Stevens, 1949.

Grotius, H. *De jure belli ac pacis* (On the law of war and peace), trans. F. W. Kelsey from 1646 ed., vol. II. Washington, D.C.: Carnegie Endowment for International Peace, 1925.

Grzybowski, K. *The Socialist Commonwealth of Nations*. New Haven: Yale University Press, 1964.
Hackworth, G. H. *Digest of International Law*, vol. V. Washington, D.C.: U.S. Government Printing Office, 1943.
Hai Fu. *Wei shih-ma i pien tao* (Why lean to one side)? Peking: Shih-chieh chih-shih she, 1951.
Halleck, H. W. *International Law*. San Francisco: H. H. Bancroft, 1861.
Hertslet, G. E. P. *Hertslet's China Treaties*, 3rd ed., vol. I. London: His Majesty's Stationery Office, 1908.
Hinckley, Frank E. "Consular Authority in China by New Treaty," *Proceedings of the American Society of International Law: Twenty-First Annual Meeting* (1927), pp. 82–87.
Hinton, Harold C. *Communist China in World Politics*. Boston: Houghton Mifflin, 1966.
Hornbeck, Stanley K. "The Most-Favored-Nation Clause," *AJIL*, 3:395–422, 619–647 (1909).
Hsia Tao-tai. *Guide to Selected Legal Sources of Mainland China*. Washington, D.C.: Library of Congress, 1967.
——"Settlement of Dual Nationality between Communist China and Other Countries," *Osteuropa–Recht* (West Germany), 11:27–38 (1965).
Hsiao, Gene T. "Communist China's Foreign Trade Contracts and Means of Settling Disputes," *Vanderbilt Law Review*, 22:503–529 (April 1969).
——"Communist China's Trade Treaties and Agreements," *Vanderbilt Law Review*, 21:623–658 (October 1968).
——"Communist Chinese Foreign Trade Organization," *Vanderbilt Law Review*, 20:303–319 (March, 1967).
Hsin Wu. "A Criticism of the Bourgeois International Law on the Question of State Territory," *KCWTYC*, 7:47–51 (1960).
Hsiung, James C. "Communist China's Conception of World Public Order," unpub. diss., Columbia University, 1967.
——*Law and Policy in China's Foreign Relations*. New York: Columbia University Press, 1971.
Huang Shun-ch'ing, Lin Hsiung-hsiang, and Kuo Hai-ming. *T'ai-wan sheng t'ung-chih kao* (Draft history of Taiwan Province), vol. X: *Kuang fu chih* (History of restoration). Taipei: T'ai-wan sheng wen-hsien wei-yuan hui, 1952.
Huang Teng Young. *The Doctrine of Rebus Sic Stantibus in International Law*. Shanghai: Commercial Press, 1935.
Huang Yü. "Such Cooperation," *People's Daily*, Dec. 31, 1956, p. 6.
Hudson, Manley O. *International Legislation*, vol. V (1929–1931). Washington, D.C.: Carnegie Endowment for International Peace, 1936.
Hyde, Charles C. *International Law, Chiefly as Interpreted and Applied by the United States*, vol. I, 2nd ed. Boston: Little, Brown, 1947.

I Hsin. "What Does Bourgeois International Law Explain about the Question of Intervention?" *KCWTYC*, 4:47–54 (1960).

The Important Documents of the First Plenary Session of the Chinese People's Political Consultative Conference. Peking: Foreign Languages Press, 1949.

"India Urged to Honor Postal Agreement with China," *SCMP*, 3043:19–20 (Aug. 20, 1963).

Information Department of the Chinese Foreign Ministry. "Chenpao Island Has Always Been Chinese Territory" (Mar. 10, 1969), *PR*, 12:14–15, no. 11 (Mar. 14, 1969).

Inspector General of Customs. *Treaties, Conventions* [etc.] *between China and Foreign States*, 2nd ed., 3 vols. Shanghai: Statistical Department, Inspectorate General of Customs, 1917.

International Affairs. Moscow: The All-Union Society "Knowledge," 1955 —.

International Court of Justice. *Reports of Judgments, Advisory Opinions and Orders 1949.* Leyden: A. W. Sijthoff, n.d.

Jackson, W. A. Douglas. *Russo-Chinese Borderlands: Zone of Peaceful Contact or Potential Conflict*? Princeton: Van Nostrand, 1962.

Jain, J. P. "The Legal Status of Formosa," *AJIL*, 57:25–45 (1963).

"*Jen-min jih-pao* Commentator Condemns U.S.-Chiang Kai-shek Illegal 'Status Agreement,'" *SCMP*, 3644:41 (Feb. 19, 1966).

Jih-pen wen-t'i wen-chien hui-pien (Collection of documents on the problem of Japan), vol. II. Peking: Shih-chieh chih-shih she, 1958.

Johnston, Douglas. "Treaty Analysis and Communist China: Preliminary Observations," *Proceedings of the American Society of International Law: Sixty-First Annual Meeting* (1967), pp. 126–134.

Johnston, Douglas, and Hungdah Chiu. *Agreements of the People's Republic of China, 1949–1967: A Calendar.* (Cambridge, Mass.: Harvard University Press, 1968).

Journal officiel de la République française: lois et decrets, Jan. 10, 1957. Paris: Imprimerie des Journaux Officiels, 1957.

Kelsen, H. *The Law of the United Nations.* London: Stevens, 1950.

King, Gillian, ed. *Documents on International Affairs, 1958.* New York and London: Oxford University Press, 1962.

Kozhevnikov, F. I., ed. *International Law* (a textbook for use in law schools), trans. by Dennis Ogden. Moscow: Foreign Languages Publishing House [1961].

——"Nekotorye voprosy teorii i praktiki mezhdunarodnogo dogovora," *Sovetskoe gosudarstvo i pravo*, 2:62–76 (1954).

K'ung Meng. "A Criticism of the Theories of Bourgeois International Law on the Subjects of International Law and the Recognition of States," *KCWTYC*, 2:44–53 (1960).

Kuo Chao. "The Names and Kinds of International Treaties," *Chi-lin jih-pao* (Kirin daily [Changchun]), April 4, 1957, p. 4.

Kuo Ch'un. *Lien-ho kuo* (The United Nations). Peking: Shih-chieh chih-shih she, 1956.

Kuo-chi t'iao-yüeh chi (International treaty series), 1934–1944; 1953–1955. Peking: Shih-chieh chih-shih ch'u pan-she, 1961.

Kuo-chi wen-t'i yen-chiu (Studies in international problems). Peking: Shih-chieh chih-shih ch'u-pan she, 1959–1960, 1964–1966.

Kuo-fu i-chiao, chien-kuo ta-kang, chung-yao hsüan-yen (Teaching of the founder of the Republic [of China], outline of national construction [and] important declarations). No publication information.

Kuo-wu yuan fa-chih chü [and] Chung-hua jen-min kung-ho kuo fa-kuei hui-pien pien-chi wei-yuan-hui pien. *Chung-hua jen-min kung-ho kuo fa-kuei hui-pien* (Collection of laws and regulations of the People's Republic of China), vols. I (1954–1955), IV (1956), VII (1958). Peking: Fa-lü ch'u-pan she, 1956–1958.

——[and] Kuo-wu yuan fa-kuei pien-tsuan wei-yuan hui pien. *Chung-hua jen-min kung-ho kuo fa-kuei hui-pien* (Collection of laws and regulations of the People's Republic of China); vol. X (1959). Peking: Fa-lü ch'u-pan she, 1960.

Kuo-wu yuan fa-kuei pien-tsuan wei-yuan hui pien. *Chung-hua jen-min kung-ho kuo fa-kuei hui-pien* (Collection of laws and regulations of the People's Republic of China), vols. XI (1960), XII (1961), XIII (1963). Peking: Fa-lü ch'u-pan she, 1960–1964.

Lall, Arthur. *How Communist China Negotiates.* New York: Columbia University Press, 1968.

League of Nations Treaty Series, vols. 9 (1922), 37 (1925), 94 (1929), 107 (1930–1931), 110 (1930–1931), 135 (1932–1933), 160 (1935), 163 (1935–1936), 205 (1944–1946). Geneva: Publication Department of the League of Nations.

Lee, Luke T. *China and International Agreements.* Leyden: A. W. Sijthoff, 1969.

——"Treaty Relations of the People's Republic of China: A Study of Compliance," *University of Pennsylvania Law Review*, 116:244–314 (1967).

Leng, Shao-chuan. "Communist China's Position on Nuclear Arms Control," *Virginia Journal of International Law*, 7:101–116 (1966).

Leng, Shao-chuan, and Hungdah Chiu. *Communist China and Selected Problems of International Law.* To be published in 1972.

"Letter of the Central Committee of the C.P.C. of February 29, 1964, to the Central Committee of the C.P.S.U.," *PR*, 7:12–18, no. 19 (May 8, 1964).

"Letter of the Central Committee of the C.P.S.U. of November 29, 1963, to the Central Committee of the C.P.C.," *PR*, 7:18–21, no. 19 (May 8, 1964).

Li, Victor H. "Legal Aspects of Trade with Communist China," *Columbia Journal of Transnational Law*, 3:57–71 (1964).

Li Pan-to. "Saar Treaty – A Dirty Deal," *People's Daily*, Jan. 2, 1957, p. 6.

Li Tieh-tseng. *Tibet: Today and Yesterday*. New York: Bookman, 1960.

Li-fa-yüan kung-pao (Gazette of the Legislative Yüan), 34th sess., no. 6 (1964); 36th sess., no. 8 (1966). Taipei: Secretariat of the Legislative Yüan, 1964, 1966.

Lin Fu-shun. *Chinese Law: Past and Present*. New York: East Asian Institute, Columbia University, 1966.

Lin Hsin. "On the System of International Law After the Second World War," *CHYYC*, 1:34–38 (1958).

MacMurray, John V. H., ed. *Treaties and Agreements with and concerning China, 1894–1919*, 2 vols. New York: Oxford University Press, 1921.

McNair, Arnold D. *The Law of Treaties*, 2nd ed. London: Oxford University Press, 1961.

——*Legal Effects of War*, 3rd ed. Cambridge, Eng.: Cambridge University Press, 1948.

Maurer, Ely. "Legal Problems Regarding Formosa and the Offshore Islands," *DSB*, 39:1005–1011, no. 1017 (Dec. 22, 1958).

Mei Ju-ao. "The Aggressor and the Law," *PC*, 5:10–14 (Mar. 1, 1955).

——"Stripping the Aggressor of Its Legal Pretext," *People's Daily*, Jan. 31, 1955, p. 3.

Millard, Thomas F. F. *The End of Extraterritoriality in China*. Shanghai: A.B.C., 1931.

Ministry of Foreign Affairs. *Treaties between the Republic of China and Foreign States 1927–1957*. Taipei: Ministry of Foreign Affairs, 1958.

Mochtar, Kusuma-atmadja. "The Sino-Indonesian Dual Nationality Treaty," in J. A. Cohen, ed., *China's Practice of International Law: Some Case Studies* (to be published in 1971).

Modzhorian, Lidiia H., and Andrei V. Sobakin. *Mezhdunarodnoe pravo v izbrannykh dokumentakh*, vol. I. Moscow: Izd-vo IMO, 1957.

Morello, Frank P. *The International Legal Status of Formosa*. The Hague: Martinus Nijhoff, 1966.

"Moscow Tries to Pull the Wool over the Eyes of Public, Says *Jen-min jih-pao*," *SCMP*, 3997:47–48 (Aug. 9, 1967).

"National Front for Liberation Is the Sole Representative of South Vietnamese People, The" (editorial), *People's Daily*, June 18, 1965, p. 1.

Notes, Memoranda and Letters Exchanged between the Governments of India and China (October 1962–January 1963), White Paper, no. VIII, and *(January 1963–July 1963), White Paper*, no. IX. New Delhi: Ministry of External Affairs, Government of India, 1963.

"NPC Standing Committee Holds 106th Meeting," *SCMP*, 3100:1 (Nov. 14, 1963).

"NPC Standing Committee Holds 119th Meeting – Decides that Chairman Liu Shao-ch'i Should Sign China-Yemen Friendship Treaty," *SCMP*, 3237:9 (June 12, 1964).

Observer, "Jordan Announces the Abrogation of the Anglo-Jordanian Treaty," *People's Daily*, Nov. 29, 1956, p. 6.

——"Why the Tripartite Treaty Does Only Harm and Brings No Benefits," *People's Daily*, Aug. 10, 1963, p. 1.

O'Connell, D. P. *International Law*, vol. I. London: Stevens, 1965.

——"The Status of Formosa and the Chinese Recognition Problem," *AJIL*, 50:405–416 (1956).

Office of Government Spokesman. *Selected Speeches and Messages of President Chiang Kai-shek, 1949–52*. Taipei: n.p., 1952.

"On China's Recognition of the 1949 Geneva Conventions," *PC*, 15:33 (Aug. 1, 1952).

"On China's Recognition of the Protocol of June 17, 1925 Prohibiting Chemical and Bacteriological Warfare," *PC*, 15:33 (Aug. 1, 1952).

"On Mao Tse-tung's Talk with a Group of Japanese Socialists" (editorial), *Pravda*, Sept. 2, 1964, p. 2.

"On the Procedure for the Ratification and Denunciation of International Treaties of the USSR," *Soviet Statutes and Decisions*, 3:56–57, no. 4, (Summer 1967).

Oppenheim, L. *International Law*, vol. I, 7th ed. by H. Lauterpacht (1948), 8th ed. Lauterpacht (1955); vol. II, 7th ed. Lauterpacht (1952). London: Longmans, Green.

Oppose the New U.S. Plots to Create "Two Chinas." Peking: Foreign Languages Press, 1962.

P'an Lang. *Meng-ku jen-min kung-ho kuo* (The Mongolian People's Republic). Peking: Ching hua yin-shu chü, 1950.

Pao Tsun-p'eng, Wu Hsiang-hsiang, and Li Ting-i. *Chung-kuo chin-tai shih lun ts'ung* (Essays on the modern history of China), 2nd ser., vol. I: *Pu p'ing teng t'iao-yüeh yü p'ing t'eng hsin yüeh* (The unequal treaties and new equal treaties). Taipei: Cheng-chung shu-chü, 1958.

Papers Relating to the Foreign Relations of the United States: 1919. Washington, D.C.: Government Printing Office, 1934.

Parry, Clive. *The Sources and Evidences of International Law*. Manchester: Manchester University Press, 1966.

"Peking Military Control Committee Requisitions Foreign Barracks in City," *NCNA*, Daily News Release no. 261 (Peking, Jan. 19, 1950), pp. 77–78.

"Peking Refuses to Cooperate," *CDSP*, 19:9, no. 29 (Aug. 9, 1967).

Peking Review. Peking, 1958 —.

People's China. Peking: Foreign Languages Press, 1950–1958.

Phillimore, Robert. *Commentaries upon International Law*. 4 Vols. 2nd ed. London: Butterworth, 1871.

Potter, B. Pitman. "Legal Aspects of the Beirut Landing," *AJIL*, 52:727–730 (1958).

"Prague's 'New Reality'" (editorial), *New York Times*, Oct. 21, 1968, p. 46.

"Premier Chou En-lai's Interview with Japanese Press and Radio," *SCMP*, 1582:21–24 (Aug. 1, 1957).

President Chiang Kai-shek: Selected Speeches and Messages in 1955. Taipei: Government Information Office, 1956.

"President Nyerere's Speech," *PR*, 11:7–9, no. 26 (June 28, 1968).

"Proclamation Defining Terms for Japanese Surrender," *DSB*, 13:137–138, no. 318 (July 29, 1945).

"Protest Against Czechoslovak Revisionist Clique's Breach of Cultural Cooperation Agreement," *PR*, 10:54, no. 32 (Aug. 4, 1967).

Pufendorf, Samuel. *De jure naturae et gentium* (On the law of nature and nations), vol. II, trans. C. H. Oldfather and W. A. Oldfather from 1688 ed. Washington, D.C.: Carnegie Endowment for International Peace, 1934.

Putney, Albert H. "The Termination of Unequal Treaties," *Proceedings of the American Society of International Law: Twenty-First Annual Meeting* (1927), pp. 87–90.

"Questions for Readjustment Submitted by China to the Peace Conference," *Chinese Social and Political Science Review*, 5a:116–161 (1920).

"Red China Quits Joint Nuclear Research Body," *China News* (Taipei), July 14, 1966.

"Refuting U.S. State Department: Chinese Statement on the Question of Exchanging Correspondents between China and the U.S.," *PR*, 3:29–31, no. 37 (Sept. 14, 1960).

"Results of Referendum in Viet-Nam," *DSB*, 33:760, no. 854 (Nov. 7, 1955).

Schwarzenberger, G. *International Law*, vol. I, 3rd ed. London: Stevens, 1957.

Selected Works of Mao Tse-tung, 4 vols. English version. Peking: Foreign Languages Press, 1961–1965.

Seyersted, Fin. *United Nations Forces in the Law of Peace and War.* Leyden: A. W. Sijthoff, 1966.

——"United Nations Forces: Some Legal Problems," *British Year Book of International Law, 1961*, 37:351–475. London and New York: Oxford University Press, 1962.

Shao Chin-fu, "The Absurd Theory of 'Two Chinas' and Principles of International Law," *KCWTYC*, 2:7–17 (1959).

Shih Hung-ping. "Hou Wai-lu Is An Experienced Anti-Communist," *People's Daily*, Nov. 22, 1966, p. 6.

Shih Sung, Yü Ta-hsin, Lu Ying-lui, and Ts'ao K'o. "An Initial Investigation into the Old Law Viewpoint in the Teaching of International Law." *CHYYC*, 4:14–16 (1958).

Shu Yuan. *Meng-ku jen-min kung-ho kuo* (The Mongolian People's Republic). Peking: Shih-chieh chih-shih ch'u-pan she, 1961.

"Sino-Burmese Agreement on Economic and Technical Cooperation," *PR*, 4:8–9, no. 2 (Jan. 13, 1961).
"Sino-Cambodian Treaty of Friendship and Mutual Non-Aggression," *PR*, 4:10, no. 19 (May 5, 1961).
"Sino-Congolese(B) Friendship Treaty," *PR*, 8:18, no. 3 (Jan. 15, 1965).
Sino-Indian Boundary Question, The, rev. ed. Peking: Foreign Languages Press, 1962.
"Sino-Indonesian Joint Communiqué on Preliminary Negotiations on Dual Nationality," *SCMP*, 957:9 (Dec. 30, 1954).
"Sino-Nepalese Treaty of Peace and Friendship, The" *PR*, 3:6–7, no. 18 (May 3, 1960).
"Sino-Tanzanian Treaty of Friendship," *PR*, 8:9, no. 9 (Feb. 26, 1965).
"Sino-Yemeni Friendship Treaty," *PR*, 7:10–11, no. 25 (June 19, 1964).
Slusser, Robert, and Jan. F. Triska. *A Calendar of Soviet Treaties, 1917–1957*. Stanford: Stanford University Press, 1959.
Snyder, Richard. *The Most-Favored-Nation Clause*. New York: King's Crown, 1948.
Sørensen, Max, ed. *Manual of Public International Law*. New York: St. Martin's, 1968.
"Soviet Notes on Border Conflict With China," *CDSP*, 21:9–13, no. 24 (July 9, 1969).
"Soviet Statement on Border Clashes Urges Negotiation," *CDSP*, 21:3–5, no. 13 (Apr. 16, 1969).
Starke, J. G. *An Introduction to International Law*, 6th ed. London: Butterworths, 1967.
"State Council Adopts Draft Sino-Nepalese Boundary Treaty at 113th Meeting," *SCMP*, 2596:1 (Oct. 11, 1961).
"State Council Holds 68th Meeting," *SCMP*, 1691:1 (Jan. 15, 1958).
"State Council Holds 146th Meeting – Approves (Draft) Treaty of Friendship Between China and Yemen," *SCMP*, 3237:9 (June 12, 1964).
"State Council Passes at 136th Plenary Session Border Treaty with Afghanistan," *SCMP*, 3088:1 (Oct. 28, 1963).
"Statement by Chinese Embassy in Indonesia on Forcible House Arrest of Chinese Consul," *PR*, 3:34–36 no. 20 (May 17, 1960).
"Statement by Chou En-lai on the Indo-China Question, June 9, 1954," *PC*, 13:5–11, supp. (July 1, 1954).
"Statement of the Government of the People's Republic of China, May 24, 1969," *PR*, 12:3–9, no. 22 (May 30, 1969).
Su-lien k'o-hsüeh yüan fa-lü yen-chiu so pien (Institute of [state and] Law of the Soviet Academy of Sciences, ed.). *Kuo-chi-fa* (International law), trans. from Kozhevnikov, ed., *Mezhdunarodnoe pravo* (Moscow: Gosiuridizdat, 1957). Peking: Shih-chieh chih-shih ch'u-pan she, 1959.
Survey of the China Mainland Press. Hong Kong: U.S. Consulate-General, 1950 —.

Ta-kung pao she jen-min shou-ts'e pien-chi wei-yüan-hui. *Jen-min shou-ts'e* (People's handbook). Peking: Hsin-hua shu-tien, 1951–1966.

Tammes, A. J. P. "Decisions of International Organs as a Source of International Law," in Academie de Droit International, *Recueil des cours*, 94:265–359 (1958). Leyden: A. W. Sijthoff, 1959.

Tan Wen-jui. "Don't Allow the Use of International Treaties as a Smoke Screen," *People's Daily*, Apr. 12, 1955, p. 4.

Taylor, Hannis. *A Treatise on International Public Law*. Chicago: Callaghan, 1901.

"Text of Sino-Ghanaian Economic and Technical Cooperation Agreement," *SCMP*, 2567:32–34 (Aug. 28, 1961).

Tibet and the Chinese People's Republic. A Report to the International Commission of Jurists by its Legal Inquiry Committee on Tibet. Geneva: International Commission of Jurists, 1960.

Ting Ku. "Firmly Maintain the Five Principles of Peaceful Coexistence," *KCWTYC*, 2:1–6 (1959).

Treaties and Agreements with and concerning China: 1919–1929. Washington, D.C.: Carnegie Endowment for International Peace, 1929.

"Treaty of Friendship," *PR*, 4:7, no. 34 (Aug. 25, 1961).

"Treaty of Friendship and Mutual Non-Aggression between the People's Republic of China and the Union of Burma," *PR*, 3:13, no. 5 (Feb. 2, 1960).

"Treaty of Friendship between the People's Republic of China and the Republic of Indonesia," *PR*, 4:11, no. 24 (June 16, 1961).

Triska, Jan F., and Robert Slusser. *The Theory, Law and Policy of Soviet Treaties*. Stanford: Stanford University Press, 1962.

Tseng Yu-hao. *The Termination of Unequal Treaties in International Law*. Shanghai: Commercial Press, 1933.

Ts'ui Shu-ch'in. *Kuo-chi fa* (International law), vol. I. Shanghai: Shang-wu yin shu kuan, 1947.

Tunkin, G. I. "O nekotorykh voprosakh mezhdunarodnogo dogovora v sviazi s Varshavskim dogovorom," *Sovetskoe gosudarstvo i pravo*, 1:98–104 (1956).

——"Parishskie Soglasheniia i mezhdunarodnoe pravo," *Sovetskoe gosudarstvo i pravo*, 2:13–22 (1955).

"UNICEF in North China Violates Shihchiachuang Understanding," *NCNA*, Daily News Release no. 244 (Jan. 1, 1950), pp. 4–6.

United Nations Documents. A/1123 (1950). A/1364 (1950). A/2361/add. 2 (1953). A/2890 (1955). A/3631 (1957). A/6309/rev. 1 (1966). A/CN. 4/101 (1956). S/1715 (1950).

United Nations International Law Commission. Draft Articles on the Law of Treaties. U.N. General Assembly: *Official Records*, 21st sess., supp. no. 9 (U.N. Doc. A/6309/rev. 1), 1966.

United Nations Treaty Series, vols. 10 (1947), 25 (1949), 41 (1949), 43 (1945), 45 (1949–50), 75 (1950), 77 (1950–51), 79 (1951), 94 (1951), 136 (1952), 177 (1953), 191 (1954), 219 (1955), 226

(1956), 227 (1956), 236 (1956), 259 (1957), 260 (1957), 263 (1957), 264 (1957), 278 (1957), 299 (1958), 313 (1958), 337 (1959), 456 (1963), 480 (1963). New York: United Nations.

"U.S. Air and Sea Forces Ordered Into Supporting Action," *DSB*, 23:5–6, no. 547 (July 3, 1950).

"U.S.-Chiang Illegally Signed 'Status Agreement' concerning U.S. Force of Aggression," *People's Daily*, Sept. 13, 1965. p. 2.

"U.S.-Chiang Kai-shek Illegal 'Status Agreement': A New Step to Make Taiwan U.S. War Base," *People's Daily*, Feb. 14, 1966, p. 2.

United States Department of State. *Treaties in Force: 1963, 1964, 1965.* Washington, D.C.: U.S. Government Printing Office, 1964–1966.

"U.S. Policy on Non-Recognition of Communist China," *DSB*, 39:385–390, no. 1002 (Sept. 8, 1958).

"United States Policy Toward Formosa," *DSB*, 22:79–81, no. 550 (Jan. 16, 1950).

"U.S., Red China Announce Measures for Return of Civilians," *DSB*, 33:456, no. 847 (Sept. 19, 1955).

"U.S. Replies to Soviet Note on Berlin," *DSB*, 40:79–89, no. 1021 (Jan. 19, 1959).

United States Statutes at Large, vols. 55, 57, 59, 61, 63. Washington, D.C.: U.S. Government Printing Office.

United States Treaties and Other International Agreements, vols. 3, 4, 6, 8, 14, 17. Washington, D.C.: U.S. Government Printing Office.

Vattel, E. *Le Droit des gens: ou principes de la loi naturelle* (The law of nations or the principles of natural law), trans. C.G. Fenwick from 1758 ed., vol. III. Washington D.C.: Carnegie Endowment for International Peace, 1916.

"Vice-Premier Chen Yi's Press Conference: China Is Determined to Make All Necessary Sacrifices for the Defeat of U.S. Imperialism," *PR*, 8:7–14, no. 41 (Oct. 8, 1965).

"Victory for the Patriotic Anti-Imperialist Struggle by Compatriots in Macao," *PR*, 10:31, no. 6 (Feb. 3, 1967).

Wai-chiao hsüeh-yuan kuo-chi fa chiao-yen-shih pien (Office of teaching and research of international law of the Institute of Diplomacy, ed.). *Kuo-chi kung-fa ts'an-k'ao wen-chien hsüan-chi* (Selected reference documents of public international law). Peking: Shih-chieh chih-shih ch'u-pan she, 1958.

Wang Fu-yen. *Kuo-chi kung-fa lun* (On the public international law), vol. I. Shanghai: Fa-hsüeh pien-i she, 1933.

Wang Shih-chieh and Hu Ching-yu. *Chung-kuo pu-p'ing teng t'iao-yüeh chih fei-ch'u* (The abolition of China's unequal treaties). Taipei: Chung-yang wen-wu kung-ying she, 1967.

Wang T'ieh-ya. *Chan-cheng yü t'iao-yüeh* (War and treaties). Chunking: Chung-kuo wen-hua fu-wu she, 1944.

Wang Yao-t'ien. *Kuo-chi mao-yi t'iao-yüeh ho hsieh-ting* (International trade treaties and agreements). Peking: Ts'ai-cheng ching-chi ch'u-pan she, 1958.

Wang Yih-wang. "What Is the Difference between Commercial Treaties and Trade Agreements," *Enlightenment Daily*, May 12, 1950, p. 3.

"Warsaw Treaty Has Become Instrument for Soviet Revisionists' Aggression Against and Enslavement of the People of Member State," *PR*, 11:8–14, no. 38 (Sept. 20, 1968).

Wei Liang. "Looking at the So-Called McMahon Line from the Angle of International Law," *KCWTYC*, 6:46–52 (1959).

——"On the Post Second World War International Treaties," *Kuo-chi t'iao-yüeh chi* (International treaty series), 1953–1955, pp. 660–690. Peking: Shih-chieh chih-shih ch'u-pan she, 1961.

West California Reporter, vol. XXXIX. St. Paul, Minn.: West, 1964.

Wheaton, Henry. *Elements of International Law*, 6th ed. Boston: Little, Brown, 1855.

Whiteman, Marjorie M. *Digest of International Law*, vol. III. Washington, D.C.: U.S. Government Printing Office, 1964.

Wolff, Christian von. *Jus gentium methodo scientifica pertratatum* (The law of nations treated according to a scientific method), vol. II, trans. Drake from 1764 ed. Washington, D.C.: Carnegie Endowment for International Peace, 1934.

Woodhead, H. G. W. *The China Year Book 1924–25*. Tientsin: Tientsin Press, n.d.

——*The China Year Book 1932*. Shanghai: North-China Daily News and Herald, n.d.

Woolsey, Theodore D. *Introduction to the Study of International Law*, 3rd ed. New York: Charles Scribner, 1871.

Wu K'un-wu. *T'iao-yüeh lun* (On treaties). Shanghai: Shang-wu yin shu kuan, 1933.

Yang Hsin and Chen Chien. "Expose and Censure the Imperialist's Fallacy concerning the Question of State Sovereignty," *CFYC*, 4:6–11 (1964).

Yearbook[s] of the International Law Commission: 1954, 1966, vol. II. U.N. Docs. A/CN.4/Ser.A/1954, 1966/Add.1. New York: United Nations, 1960, 1970.

Yearbook of the United Nations, 1950. New York: Columbia University Press, with the United Nations, 1951.

Ying T'ao. "A Criticism of Bourgeois International Law concerning the Question of State Sovereignty," *KCWTYC*, 3:47–52 (1960).

——"Recognize the True Face of Bourgeois International Law from a Few Basic Concepts," *KCWTYC*, 1:47–51 (1960).

Yokota Kisaburo. *Kokusai hogaku* (The study of international law), vol. I. Tokyo: Yuhikaku, 1955.

Young, Kenneth T. *Negotiating with the Chinese Communists: The United States Experience, 1953–1967*. New York: McGraw-Hill, 1968.

Yü Fan. "Speaking about the Relationship between China and the Tibetan Region from the Viewpoint of Sovereignty and Suzerainty," *People's Daily*, June 5, 1959, p. 7.

Zadorozhny, G. "Peking Opposes Cooperation," *Izvestia*, July 19, 1967, p. 2.

Glossary

abrogation; fei-ch'u 廢除
accession; chia-ju 加入
accord; hsieh-i 協議
agreement; hsieh-ting 協定
alliance; t'ung-meng 同盟
announcement; sheng-ming 聲明
approval; ho-chun 核准
arbitration; chung-ts'ai 仲裁

boundary; pien-chieh 邊界

chairman; chu-hsi 主席
charter; hsien-chang 憲章
collective self-defense; chi-t'i tzu-wei 集體自衛
commerce; t'ung-shang 通商
company; kung-szu 公司
consular treaty; ling-shih t'iao-yüeh 領事條約
contract; ho-t'ung 合同
contracting parties; t'i-yüeh shuang-fang 締約雙方
convention; kung-yüeh, chuan-yüeh 公約; 專約
corporation; kung-szu 公司
credentials; ch'üan-ch'üan cheng-shu 全權證書
cultural cooperation; wen-hua ho-tso 文化合作

declaration; hsüan-yen 宣言
diplomatic channel; wai-chiao t'u-ching 外交途徑
dispute; cheng-tuan 爭端

economic aid; ching-chi yüan-chu 經濟援助
economic and technical cooperation; ching-chi chi-shu ho-tso 經濟技術合作
exchange of goods; chiao-huan huo-wu; huan-huo; huo-wu chiao-huan 交換貨物; 換貨; 貨物交換
exchange of notes; huan-wen 換文
executive organ; chih-hsing chi-kou 執行機構

friendship; yu-hao 友好
foreign minister, see minister of foreign affairs
full power; ch'üan-ch'üan 全權

general conditions for delivery of goods; chiao-huo kung-t'ung t'iao-chien 交貨共同條件
goods; huo-wu 貨物

initialing; ts'ao-ch'ien 草簽
instrument of ratification; p'i-chun shu 批准書
international law; kuo-chi fa 國際法
interpretation; chieh-shih 解釋
intervention; kan-she 干涉

joint arrangement; kung-t'ung pan-fa 共同辦法
joint communiqué; lien-ho kung-pao 聯合公報
joint statement; lien-ho sheng-ming; kung-t'ung sheng-ming 聯合聲明; 共同聲明

law of treaties; t'iao-yüeh fa 條約法
loan; tai-k'uan 貸款
local authority; ti-fang tang-chü 地方當局

memorandum; pei-wang-lu 備忘錄
minister of foreign affairs; wai-chiao pu-chang 外交部長

ministry of foreign affairs; wai-chiao pu 外交部
minute; chi-lu-shu; chi-yao 記錄書; 紀要
mixed commission; hun-ho wei-yüan hui 混合委員會
most-favored nation; tsui-hui-kuo 最惠國
multilateral treaty; to-pien t'iao-yüeh 多邊條約
municipal law; kuo-nei fa 國內法
mutual assistance; hu-chu 互助
mutual nonaggression; hu pu ch'in-fan 互不侵犯
mutual supply of goods; hu-hsiang kung-ying huo-wu 互相供應貨物

National People's Congress; ch'üan-kuo jen-min tai-piao ta-hui 全國人民代表大會
nationality; kuo-chi 國籍
navigation; hang-hai 航海
negotiation; hsieh-shang; t'an-p'an 協商; 談判
neutrality; chung-li 中立
note; chao-hui 照會

object; mu-ti; k'o-t'i 目的; 客體

pacta sunt servanda (treaties are to be kept); t'iao-yüeh pi-hsü tsun-shou 條約必須遵守
pacta tertiis nec nocent nec prosunt (treaties do not impose any burden, nor confer any benefits, upon third states); t'i-yüeh shuang-fang chien ti hsieh-ting pu-neng chü-shu ti-san-kuo 締約双方間的協定不能拘束第三國
payment; chih-fu; fu-k'uan 支付; 付款
peaceful coexistence; ho-p'ing kung-ch'u 和平共處
plan; chi-hua 計劃
plenipotentiary; ch'üan-ch'üan tai-piao 全權代表
premier; tsung-li 總理
press communiqué; hsin-wen kung-pao 新聞公報
proclamation; kung-kao 公告
protected state; pei pao-hu kuo 被保護國
protocol; i-ting-shu 議定書
purpose; mu-ti; tsung-chih 目的; 宗旨

ratification; p'i-chun 批准
rebus sic stantibus (vital change of circumstances); ch'ing-shih pien-ch'ien 情勢變遷
recognition; ch'eng-jen 承認
registration; teng-chi 登記
reservation; pao-liu 保留

scientific and technical cooperation; k'o-hsüeh ho chi-shu ho-tso 科學和技術合作
signature; ch'ien-tzu 簽字
source; yüan-yüan 淵源
sovereignty; chu-ch'üan 主權
standing committee; ch'ang wu wei-yüan hui 常務委員會
State Council; kuo-wu yüan 國務院
suzerainty; tsung chu-ch'üan 宗主權

talk; hui-t'an 會談
trade; mao-i 貿易
treaty; t'iao-yüeh 條約

unequal treaty; pu-p'ing-teng t'iao-yüeh 不平等條約
United Nations; lien-ho-kuo 聯合國

vassal state; fu-yung kuo 附庸國
violation; wei-fan; p'o-huai 違反; 破壞
vital change of circumstances, see *rebus sic stantibus*

war; chan-cheng 戰爭

Chinese- and Japanese-Language Books, Periodicals, and Newspapers

Cheng-fa yen-chiu 政法研究. Peking: Chung-kuo cheng-chih fa-lü hsüeh-hui 中國政治法律學會, 1954–1966.

Cheng-ta fa-hsüeh p'ing-lun 政大法學評論. Taipei: Cheng-ta fa-lü hsi 政大法律系, 1969——.

Chi-lin jih-pao 吉林日報.

Chiang Ching-kuo 蔣經國. *Fu-chung chih-yüan* 負重致遠. Taipei: Yu-shih shu-tien 幼獅書店, 1963.

Chiao-hsüeh yü yen-chiu 教學與研究. Peking: Chung-kuo jen-min ta-hsüeh 中國人民大學, 1957–1959.

Ch'ien T'ai 錢泰. *Chung-kuo pu-p'ing-teng t'iao-yüeh chih yüan-ch'i chi ch'i fei-ch'u chih ching-kuo* 中國不平等條約之緣起及其廢除之經過. Taipei: Kuo-fang yen-chiu yüan 國防研究院, 1961.

Ch'in Tzu-ch'ing 秦子青. *Lun Mei Chiang ch'in-lüeh t'iao-yüeh* 論美蔣侵略條約. Peking: Shih-chieh chih-shih she 世界知識社, 1955.

Chung-hua jen-min kung-ho kuo tui-wai kuan-hsi wen-chien chi 中華人民共和國對外關係文件集. Peking: Shih-chieh chih-shih ch'u-pan she 世界知識出版社, 1957–1965.

Chung-hua jen-min kung-ho kuo wai-chiao pu pien 中華人民共和國外交部編. *Chung-hua jen-min kung-ho kuo t'iao-yüeh chi* 中華人民共和國條約集. Peking: vols. I-X, Fa-lü ch'u-pan she 法律出版社; vols. XI-XIII, Shih-chieh chih-shih ch'u-pan she 世界知識出版社; 1963–1965.

——— *Chung-hua jen-min kung-ho kuo yu-hao t'iao-yüeh hui-pien* 中華人民共和國友好條約匯編. Peking: Shih-chieh chih-shih ch'u-pan she 世界知識出版社, 1965.

——— *I-chiu-szu-chiu nien pa-yüeh shih-erh jih jih-nei-wa kung-yüeh* 一九四九年八月十二日日內瓦公約. Peking: Fa-lü ch'u-pan she 法律出版社, 1958.

Chung-hua min-kuo k'ai-kuo wu-shih nien wen-hsien pien-ch'uan wei-yuan hui pien 中華民國開國五十年文獻編纂委員會編. *Chung-hua min-kuo k'ai-kuo wu-shih nien wen-hsien* 中華民國開國五十年文獻, 1st ser., vol. VII, pt. 1: *Ch'ing-t'ing chih kai-ke yü fan-tung* 淸廷之改革與反動. Taipei: Chung-hua min-kuo k'ai-kuo wu-shih nien wen-hsien pien-ch'uan wei-yuan hui 中華民國開國五十年文獻編纂委員會, 1965.

——— *Kung-fei huo-kuo shih-liao hui-pien* 共匪禍國史料彙編, vol. I. Taipei: Chung-hua min-kuo k'ai-kuo wu-shih nien wen-hsien pien-ch'uan wei-yuan hui 中華民國開國五十年文獻編纂委員會, 1964.

Chung-shan ch'üan-shu 中山全書, vol. II. Shanghai: Ta-hua shu-chü 大華書局, 1927.

Chung-yang jen-min cheng-fu fa-chih wei-yuan hui pien 中央人民政府法制委員會編. *Chung-yang jen-min cheng-fu fa-ling hui pien* 中央人民政府法令彙編, vol. I (1949–1950). Peking: Hsin-hua shu-tien 新華書店, 1952.

Fa-hsüeh 法學. Shanghai: Shang-hai fa-hsüeh hui 上海法學會 [and] Hua-tung cheng-fa hsüeh-yüan 華東政法學院, 1957–1959.

Hai Fu 海父. *Wei shih-ma i pien tao?* 爲什麽一邊倒? Peking: Shih-chieh chih-shih she 世界知識社, 1951.

Hua-tung cheng-fa hsüeh-pao 華東政法學報. Shanghai: Hua-tung cheng-fa hsüeh-yüan 華東政法學院, 1956.

Huang Shun-ch'ing 黃純青, Lin Hsiung-hsiang 林熊祥, and Kuo Hai-ming 郭海鳴. *T'ai-wan sheng t'ung chih kao* 台灣省通志稿, vol. X: *Kuang-fu chih* 光復志. Taipei: T'ai-wan sheng wen-hsien wei-yuan hui 台灣省文獻委員會, 1952.

Jen-min jih-pao 人民日報.

Jen-wu 人物. Hong Kong: Jen-wu tsa-chih she 人物雜誌社, 1966——.

Jih-pen wen-t'i wen-chien hui-pien 日本問題文件彙編, vol. I. Peking: Shih-chieh chih-shih she 世界知識社, 1955.

Kuang-ming jih-pao 光明日報.

Kuo Ch'ün 郭群. *Lien-ho kuo* 聯合國. Peking: Shih-chieh chih-shih she 世界知識社, 1956.

Kuo-chi t'iao-yüeh chi 國際條約集, 1934–1944, 1953–1955. Peking: Shih-chieh chih-shih ch'u-pan she 世界知識出版社, 1961.

Kuo-chi wen-t'i yen-chiu 國際問題研究. Peking: Shih-chieh chih-shih ch'u-pan she 世界知識出版社, 1959–1960, 1964–1966.

Kuo-fu i-chiao, chien-kuo ta-kang, chung-yao hsüan-yen 國父遺教, 建國大綱, 重要宣言. No publication information.

Kuo-wu yüan fa-chih chü 國務院法制局 and Chung-hua jen-min kung-ho kuo fa-kuei hui-pien pien-chi wei-yuan hui pien 中華人民共和國法規彙編編輯委員會編. *Chung-hua jen-min kung-ho kuo fa-kuei hui-pien* 中華人民共和國法規彙編, vols. I (1954–1955), IV (1956), VII (1958). Peking: Fa-lü ch'u-pan she 法律出版社, 1956–1958.

——— and Kuo-wu yüan fa-kuei pien-tsuan wei-yuan hui pien 國務院法規編纂委員會編. *Chung-hua jen-min kung-ho kuo fa-kuei hui-pien* 中華人民共和國法規彙編, vol. X (1959). Peking: Fa-lü ch'u-pan she 法律出版社, 1960.

Kuo-wu yüan fa-kuei pien-tsuan wei-yuan hui pien 國務院法規編纂委員會編. *Chung-hua jen-min kung-ho kuo fa-kuei hui-pien* 中華人民共和國法規彙編, vols. XI (1960), XII (1961), XIII (1963). Peking: Fa-lü ch'u-pan she 法律出版社, 1960–1964.

Li-fa-yüan kung-pao 立法院公報, 34th sess., no. 6 (1964); 36th sess., no. 8 (1966). Taipei: Li-fa-yüan pi-shu ch'u 立法院秘書處.

P'an Lang 潘朗. *Meng-ku jen-min kung-ho kuo* 蒙古人民共和國. Peking: Ching hua yin-shu chü 京華印書局, 1950.

Pao Tsun-p'eng 包遵彭, Wu Hsiang-hsiang 吳相湘, and Li Ting-i 李定一. *Chung-kuo chin-tai shih lun ts'ung* 中國近代史論叢, 2nd ser., vol. I: *Pu-p'ing-teng t'iao-yüeh yü p'ing-teng hsin yüeh* 不平等條約與平等新約. Taipei: Cheng-chung shu-chü 正中書局, 1958.

Peking Review 北京評論. Peking: 1958——.

People's China 人民中國. Peking: Foreign Languages Press, 1950–1958.

Sekai shūhō 世界週報.

Shih-chieh chih-shih 世界知識. Peking: Shih-chieh chih-shih ch'u-pan she 世界知識出版社, 1955.

Shu Yuan 樞原. *Meng-ku jen-min kung-ho kuo* 蒙古人民共和國. Peking: Shih-chieh chih-shih ch'u-pan she 世界知識出版社, 1961.

Ta-kung pao she jen-min shou-ts'e pien-chi wei-yüan-hui 大公報社人民手冊編輯委員會. *Jen-min shou-ts'e* 人民手冊. Peking: Hsin-hua shu-tien 新華書店, 1951–1966.

Ts'ui Shu-ch'in 崔書琴. *Kuo-chi fa* 國際法, vol. I. Shanghai: Shang-wu yin shu kuan 商務印書館, 1947.

Wai-chiao hsüeh-yüan kuo-chi fa chiao-yen-shih pien 外交學院國際法教研室編. *Kuo-chi kung-fa ts'an-k'ao wen-chien hsuan-chi* 國際公法參考文件選輯. Peking: Shih-chieh chih-shih ch'u-pan she 世界知識出版社, 1958.

Wang Fu-yen 汪馥炎. Kuo-chi kung-fa lun 國際公法論, vol. I. Shanghai: Fa-hsüeh pien-i she 法學編譯社, 1933.

Wang Shih-chieh 王世杰 and Hu Ch'ing-yü 胡慶育. *Chung-kuo pu-p'ing-teng t'iao-yüeh chih fei-ch'u* 中國不平等條約之廢除. Taipei: Chung-yang wen-wu kung-ying she 中央文物供應社, 1967.

Wang T'ieh-ya 王鐵崖. *Chan-cheng yü t'iao-yüeh* 戰爭與條約. Chungking: Chung-kuo wen-hua fu-wu she 中國文化服務社, 1944.

Wang Yao-t'ien 汪堯田. *Kuo-chi mao-i t'iao-yüeh ho hsieh-ting* 國際貿易條約和協定. Peking: Ts'ai-cheng ching-chi ch'u-pan she 財政經濟出版社, 1958.

Wu K'un-wu 吳昆吾. T'iao-yüeh lun 條約論. Shanghai: Shang-wu yin shu kuan 商務印書館, 1933.

Yokota Kisaburō 横田喜三郎. *Kokusai hōgaku* 國際法學, vol. I. Tokyo: Yuhi-kaku 有斐閣, 1955.

Chinese-Language Articles

Chao Yüeh 眺岳. "A Preliminary Criticism of Bourgeois International Law" 對資產階級國際法的初步批判, *KCWTYC*, 3: 1–9 (1959).

Ch'en T'i-ch'iang 陳體强. "The Hungarian Incident and the Principle of Non-Intervention" 匈牙利事件與不干涉原則, *Enlightenment Daily*, Apr. 5, 1957, p. 1.

——— "The Illegality of Atomic Weapons from the Viewpoint of International Law" 從國際法看原子武器的非法性, *Shih-chieh chih-shih* (World knowledge), 4: 11–12 (Feb. 20, 1955).

——— "Sovereignty of Taiwan Belongs to China" 台灣的主權屬於中國, *People's Daily*, Feb. 8, 1955, p. 4.

Ch'ien Szu 千駟. "A Criticism of the Views of Bourgeois International Law on the Question of the Population" 批判資產階級國際法在居民問題上的主張, *KCWTYC*, 5: 40–49 (1960).

Chiu Hungdah 丘宏達. "The Attitude of Chinese Communists toward the 'Question of Taiwan'" 中共対「台灣問題」態度的研究, *Jen-wu* (Character magazine [Hong Kong]), 9: 6–8, 38–39 (December 1967).

——— "A Comparative Study of the Chinese and Western Position on the Problem of Unequal Treaties" 中國與西方關於不平等條約問題的比較研究, *Cheng-ta fa-hsüeh p'ing-lun* (Chengchi law review) (December 1969), pp. 1–9.

Ch'iu Jih-ch'ing 丘日慶. "Further Discussion of the System of International Law at the Present Stage" 再論現階段國際法的體系, *FH*, 3: 40–43 (1958).

Chou Fu-lun 周福倫. "On the Nature of Modern International Law" 試論現代國際法的性質, *CHYYC*, 3: 52–56 (1958).

Chou Keng-sheng 周鯁生. "China's Legitimate Rights in the United Nations Must Be Restored" 中國在聯合國的合法權利必須恢復, *People's Daily*, Dec. 5, 1961, p. 5.

——— "Don't Allow American and British Aggressors to Intervene in the Internal Affairs of Other States" 不容許美英侵略者干涉他國內政, *CFYC*, 4: 3–4 (1958).

——— "Looking at the West Berlin Question from the Angle of International Law" 從國際法的角度看西柏林問題, *KCWTYC*, 1: 40–45 (1959).

——— "The Persecution of Chinese Personnel by Brazilian Coup d'Etat Authority Is a Serious International Illegal Act" 巴西政變當局迫害中國人員是嚴重的國際不法行爲, *People's Daily*, Apr. 24, 1964, p. 4.

——— "The Principles of Peaceful Coexistence from the Viewpoint of International Law" 從國際法論和平共處的原則, *CFYC*, 6: 37–41 (1955).

Chou Tze-ya 周子亞. "Talks on the Question of Suez Canal" 談蘇伊士運河問題, *Hua-tung cheng-fa hsüeh-pao* (East China journal of political science and law), 3: 11, 35–38 (1956).

Chu Li-sun 朱荔蓀. "The Use of Atomic and Hydrogen Weapons Is the Most Serious Criminal Act in Violation of International Law" 使用原子武器和氫武器是最嚴重的違反國際法的罪行, *CFYC*, 4: 30–33 (1955).

——— "Another Crime of the U.S.-Chiang Kai-shek Conspiracy" 美蔣勾結的又一罪行, *People's Daily*, Feb. 19, 1966, p. 1.

——— "Insidious U.S.-Soviet Collaboration Policy" 美蘇合謀的陰險政策, *People's Daily*, July 23, 1967, p. 4.

Commentator 評論員. "A Nuclear Fraud Jointly Hatched by the United States and the Soviet Union" 美蘇合謀的核騙局, *People's Daily*, June 13, 1968, p. 5.

——— "The Soviet Revisionist Renegades Try to Pull the Wool Over the Eyes of the Public" 蘇修叛徒的障眼法, *People's Daily*, Aug. 5, 1967, p. 6.

"Courageous and Resolute Revolutionary Action" 英勇果敢的革命行動 (editorial), *People's Daily*, Sept. 20, 1968, p. 1.

Fu Chu 傅鑄. "American Imperialism's Use of Poisonous Gas in South Vietnam is a War Crime in Flagrant Violation of International Law" 美帝在南越使用毒氣是嚴重違反國際法的戰爭罪行, *People's Daily*, Apr. 3, 1965, p. 3.

Hsin Wu 欣梧. "A Criticism of the Bourgeois International Law on the Question of State Territory" 對資產階級國際法關於國家領土問題的批判, *KCWTYC*, 7: 47–51 (1960).

Huang Yü 黃予. "Such Cooperation" 如此合作, *People's Daily*, Dec. 31, 1956, p. 6.

I Hsin 一心. "What Does Bourgeois International Law Explain about the Question of Intervention" 資產階級國際法在干涉問題上說明了什麼, *KCWTYC*, 4: 47–54 (1960).

K'ung Meng 孔夢. "A Criticism of the Theories of Bourgeois International Law on the Subjects of International Law and the Recognition of States" 對資產階級國際法關於國際法主體和國家承認的理論的批判, *KCWTYC*, 2: 44–53 (1960).

Kuo Chao 柯召. "The Names and Kinds of International Treaties" 國際條約的名稱和種類, *Chi-lin jih-pao* (Kirin daily), Apr. 4, 1957, p. 4.

Li Pan-to 黎邦鐸. "Saar Treaty—A Dirty Deal" 薩爾條約——宗骯髒的交易, *People's Daily*, Jan. 2, 1957, p. 6.

Lin Hsin 林欣. "On the System of International Law after the Second World War" 論第二次世界大戰後的國際法體系, *CHYYC*, 1: 34–38 (1958).

Mei Ju-ao 梅汝璈. "Stripping the Aggressor of Its Legal Pretext" 剝去侵略者的法律外衣, *People's Daily*, Jan. 31, 1955, p. 3.

"The National Front for Liberation Is the Sole Representative of the South Vietnamese People" 越南南方民族解放陣綫是南越人民的唯一代表 (editorial), *People's Daily*, June 20, 1965, p. 1.

Observer 觀察家. "Jordan Announces the Abrogation of the Anglo-Jordanian Treaty" 約旦宣佈廢除英約條約, *People's Daily*, Nov. 29, 1956, p. 6.

——— "Why the Tripartite Treaty Does Only Harm and Brings No Benefits" 爲什麼三國條約有百弊而無一利, *People's Daily*, Aug. 10, 1963, p. 1.

Shao Chin-fu 邵今甫. "The Absurd Theory of 'Two Chinas' and Principles of

International Law" 「兩個中國」謬論和國際法原則, *KCWTYC*, 2: 7–17 (1959).

Shih Hung-ping 史紅兵. "Hou Wai-lu Is an Experienced Anti-Communist" 侯外盧是反共老手, *People's Daily*, Nov. 22, 1966, p. 6.

Shih Sung 史松, Yü Ta-hsin 俞大鑫, Lu Ying-hui 盧瑩輝, and Ts'ao K'o 曹柯. "An Initial Investigation into the Old Law Viewpoint in the Teaching of International Law" 對「國際公法」教學中舊法觀點的初步檢查, *CHYYC*, 4: 14–16 (1958).

T'an Wen-jui 譚文瑞. "Don't Allow the Use of International Treaties as a Smoke Screen" 不許把國際條約當作烟幕, *People's Daily*, Apr. 12, 1955, p. 4.

Ting Ku 丁谷. "Firmly Maintain the Five Principles of Peaceful Coexistence" 堅決維護和平共處的五項原則, *KCWTYC*, 2: 1–6 (1959).

"U.S.-Chiang Illegally Signed 'Status Agreement' Concerning U.S. Force of Aggression" 美蔣非法簽訂侵台美軍「地位協定」, *People's Daily*, Sept. 13, 1965, p. 2.

Wang Yih-wang 王翌望. "What Are the Differences between Commercial Treaties and Trade Agreements?" 商約和貿易協定有什麼分別? *Enlightenment Daily*, May 12, 1950, p. 3.

Wei Liang 韋良. "Looking at the So-called McMahon Line from the Angle of International Law" 從國際法角度看所謂麥克馬洪綫問題, *KCWTYC*, 6: 47–52 (1959).

——— "On the Post Second World War International Treaties" 略論第二次世界大戰後的國際條約, *Kuo-chi t'iao-yüeh chi* (International treaty series), 1953–1955, pp. 660–690. Peking: Shih-chieh chih-shih ch'u pan she, 1961.

Yang Hsin 楊欣 and Ch'en Chien 陳健. "Expose and Censure the Imperialist's Fallacy concerning the Question of State Sovereignty" 揭露和批判帝國主義者關於國家主權問題的謬論, *CFYC*, 4: 6–11 (1964).

Ying T'ao 英濤. "A Criticism of Bourgeois International Law concerning the Question of State Sovereignty" 對資產階級國際法關于國家主權問題的批判, *KCWTYC*, 3: 47–52 (1960).

——— "Recognize the True Face of Bourgeois International Law from a Few Basic Concepts" 從幾個基本概念認識資產階級國際法的眞面目, *KCWTYC*, 1: 47–51 (1960).

Yü Fan 于蕃. "Speaking about the Relationship between China and the Tibetan Region from the Viewpoint of Sovereignty and Suzerainty" 從主權和宗主權談到中國對西藏地方的關係, *People's Daily*, June 5, 1959, p. 7.

Index